# 1089 and All That

# 1089 and All That

## A Journey into Mathematics

David Acheson
*Jesus College, Oxford*

OXFORD
UNIVERSITY PRESS

# OXFORD
UNIVERSITY PRESS

Great Clarendon Street, Oxford OX2 6DP

Oxford University Press is a department of the University of Oxford.
It furthers the University's objective of excellence in research, scholarship,
and education by publishing worldwide in

Oxford New York

Auckland Bangkok Buenos Aires Cape Town Chennai
Dar es Salaam Delhi Hong Kong Istanbul Karachi Kolkata
Kuala Lumpur Madrid Melbourne Mexico City Mumbai Nairobi
São Paulo Shanghai Taipei Tokyo Toronto

Oxford is a registered trade mark of Oxford University Press
in the UK and in certain other countries

Published in the United States
by Oxford University Press Inc., New York

First published 2002
Reprinted 2003

A catalogue record for this title is available from the British Library

Cataloging in Publication Data
(Data available)
ISBN 0 19 851623 1

10 9 8 7 6 5 4 3 2 1

Printed in Great Britain
on acid-free paper by Biddles Ltd, Guildford & King's Lynn

# *Contents*

| | | |
|---|---|---|
| 1 | 1089 and All That | *1* |
| 2 | 'In Love with Geometrie' | *9* |
| 3 | But . . . that's Absurd . . . | *19* |
| 4 | The Trouble with Algebra | *29* |
| 5 | The Heavens in Motion | *41* |
| 6 | All Change! | *53* |
| 7 | On Being as Small as Possible | *61* |
| 8 | 'Are We Nearly There?' | *73* |
| 9 | A Brief History of $\pi$ | *83* |
| 10 | Good Vibrations | *93* |
| 11 | Great Mistakes | *103* |
| 12 | What is the Secret of All Life? | *113* |
| 13 | $e = 2.718 \ldots$ | *123* |
| 14 | Chaos and Catastrophe | *135* |
| 15 | Not Quite the Indian Rope Trick | *147* |
| 16 | Real or Imaginary? | *159* |
| | Suggestions for Further Reading | *171* |
| | The *1089 and All That* Website | *172* |
| | Acknowledgments | *172* |
| | Picture Credits | *173* |
| | Index | *175* |

TRICK No 4

CHAPTER ONE

# *1089 and All That*

Think of a three-figure number.

Any three-figure number will do, so long as the first and last figures differ by 2 or more.

Now reverse it, and subtract the smaller number from the larger. So, for example,

$$782 - 287 = 495.$$

Finally, reverse the new three-figure number, and add:

$$495 + 594 = 1089.$$

At the end of this procedure, then, we have a final answer of 1089, though we have to expect, surely, that this final answer will depend on which three-figure number we start with.

But it *doesn't*.

The final answer always turns out to be 1089.

———

As I remember, the '1089 trick' was the first piece of mathematics that really impressed me, and I came across it at the age of ten, in the *I-SPY Annual* for 1956.

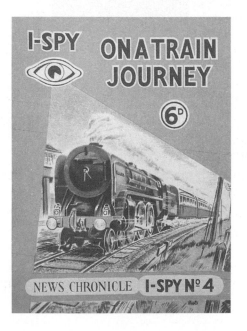

This was a book for children, published by a well-known British newspaper of the time, and it contained a mixture of adventure stories and more educational articles with titles like 'Pond Life'.

But my favourite bit, by a long way, was

**ABRACADABRA!**
*and Uncle Jack turns you into a conjuror*

A TRICK WITH NUMBERS
The conjuror writes a number on the blank side of the slate he is holding. A friend is asked to write a number of three different figures on a piece of paper. He must then turn this number round and take the smaller number from the larger and finally turn this number round and add it to the result of the subtraction.

When this has been done, the conjuror turns the slate round and shows that he has written the final number 1089.
SECRET
The number arrived at in this trick is always 1089.

TRICK No 4

There were other conjuring tricks as well, including 'The Vanishing Glass of Water' and 'Reading the Mind', but somehow it was '1089' that really caught my attention.

It was the element of mystery and surprise, I think, that put this result into a different league from some of the work we were doing in school.

© Glen Baxter

Now, I'm not saying that I didn't enjoy 'sums', and other bits of elementary mathematics, for I most certainly did. But if I tell you, for instance, that a typical homework problem at the time went something like this:

> A and B can fill a cistern in 4 hours. A and C can fill the same cistern in 5 hours. B can fill twice as fast as C. Find how long C would take to fill the cistern, working alone.*

I think you will understand why the '1089' trick made such an impression.

---

And now, over 40 years later, it seems to me that these same elements of mystery and surprise run through a great deal of mathematics at its best. Some of the first-rate theorems and results really do generate a sense of *wonder*.

I hope to show something of this as we go through the book, and I hope to show, too, how there is much pleasure to be had, from time to time, in the actual deductive arguments by which those theorems and results are *proved*.

In addition to all this, we shall take in several remarkable applications of mathematics to science and nature.

* C would in fact take 20 hours to fill the cistern, poor devil.

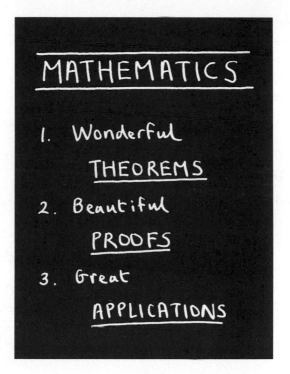

So, whether you are very young or very old, or some-where in between; whether you are at school or at university, or neither; whether you have a pen in your hand, or a gin and tonic . . . we are about to go on a journey.

Along the way we shall be taking in some of the most important ideas of mathematics, and something of their history.

We will be going, in short, from first steps to the frontiers, and in order to keep track of 'the big picture' while we're doing all this, we will be going along quite fast.

If we imagine that we are on a train, for instance, then it will be the *Mathematics Express* . . .

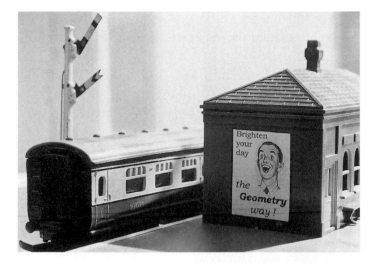

CHAPTER TWO

# *'In Love with Geometrie'*

One of the best-documented examples of someone being really surprised by mathematics is to be found in the following anecdote about the philosopher Thomas Hobbes (1588–1679):

> He was 40 yeares old before he looked on Geometry; which happened accidentally. Being in a Gentleman's Library, Euclid's Elements lay open, and 'twas the *47 El. libri 1*. He read the Proposition. *By G –*, sayd he (he would now and then sweare an emphaticall Oath by way of emphasis) *this is impossible!*

Here, then, is an example of mathematics at its best, for Hobbes found the result so stunning that he couldn't quite believe it.

The result in question was, in fact, none other than *Pythagoras' theorem*: if $a$, $b$ and $c$ are the sides of a right-angled triangle, and $c$ is the longest side, then $a^2 + b^2 = c^2$.

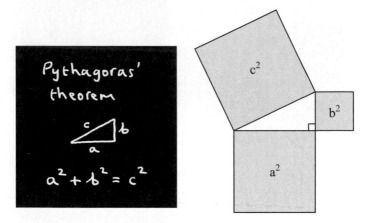

And Hobbes didn't just take somebody's word for this; he read a *proof*. It was this proof, as much as any-thing else, that made him

   . . . in love with Geometrie.

So we too will now prove Pythagoras' theorem.

———

I can see, of course, that this might prompt the ques-tion: *why bother?* After all, the theorem has been around for over 2000 years. Everybody knows Pythagoras' theorem. Surely, if it were not true, if there were anything wrong with it, *somebody would have noticed by now*.

In mathematics, however, this kind of argument is virtually worthless.

And in any case, the following delightfully simple proof of Pythagoras' theorem is almost fun.

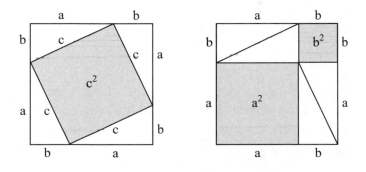

Take a square of side $a + b$ and place 4 copies of the original right-angled triangle within it, as shown. This leaves a square area $c^2$. Now think of the triangles as white tiles on a dark floor, and move three of them so that they occupy the new positions indicated. The floor area *not* occupied by triangles is now $a^2 + b^2$, yet must be the same as before.

So $a^2 + b^2 = c^2$.

Two special cases of Pythagoras' theorem are of particular interest.

One is when the smaller angles of the right-angled triangle are both 45°:

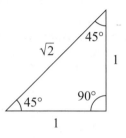

If the two shorter sides are of length 1, say, then the length of the longest side is √2, and this is one common way in which the square root of 2 arises in mathematics.

Another commonly-occurring special case is when the two smaller angles are 30° and 60°:

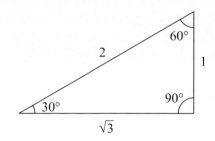

But these *are* just special cases. The real power and importance of Pythagoras' theorem lies in its *generality*; it is equally true whether the right-angled triangle in question is short and fat or long and thin.

And we know this not because Professor X – who is supposedly a world expert – assures us that it is, but because we have seen it for ourselves.

***

If Pythagoras' theorem is the most well-known result in the whole of geometry, then the next best known must surely be the formulae for the circumference and area of a circle of radius *r*:

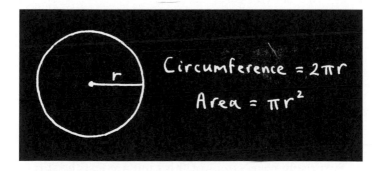

Circumference $= 2\pi r$

Area $= \pi r^2$

And this is the way in which the special number

$$\pi = 3.14159\ldots$$

first enters mathematics. In 'elementary' mathematics, $\pi$ is all about circles.

Imagine the surprise, then, in the mid-seventeenth century, when mathematicians found π cropping up in all sorts of places that had, apparently, nothing to do with circles at all.

One of the most famous results of this kind is an extra-ordinary connection between π and the *odd numbers*:

$$\frac{\pi}{4} = 1 - \frac{1}{3} + \frac{1}{5} - \frac{1}{7} + \cdots$$

Here, as the dots indicate, we are meant to keep on adding and subtracting the fractions on the right-hand side *for ever*. First, then, it is not at all obvious that the 'sum' in question settles down to any definite value at all.

But, even given that it does, why should that value be $\frac{\pi}{4}$? What on earth have circles got to do with the odd numbers 1, 3, 5, 7, . . .?

Surprising *connections* of this kind are just the sort of thing that get mathematicians really excited.

Today, the whole subject of geometry extends way beyond the world of right-angled triangles, circles and so on. There are even branches of the subject in which the ideas of length, angle and area don't really feature at all.

One of these is *topology* – a sort of rubber-sheet geometry – where a recurring question is whether some geometric object can be deformed 'smoothly' into another one.

Questions of this kind can be very demanding, and even counter-intuitive.

Look, for example, at the two geometric objects below, and ask yourself if the one on the left can be deformed smoothly into the one on the right.

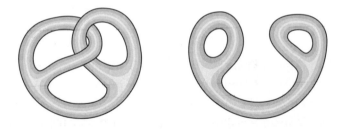

Imagine, if you will, that the object is made of some very elastic material, so that you can stretch or squash it as much as you like.

Is it possible, then, to deform the object – without cutting or tearing it – into its 'unlinked' version?

Well, actually . . . it *is*:

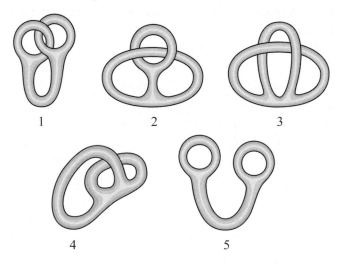

1    2    3

4    5

We end this chapter, though, by returning to one of the most important legacies of ancient Greek geometry: the concept of *proof*.

One reason for emphasizing this idea so early in the book is that it is all too easy in mathematics to jump to the wrong conclusion.

And it is particularly dangerous to jump to some general conclusion on the basis of a few special cases.

Here's an example. Take a circle, mark 2 points on the circumference and join them by a straight line. This divides the circle into 2 regions.

Now mark 3 points on the circumference instead, and join each point to *all* the others by straight lines. We get 4 regions.

If we do the same thing with 4 points, we get 8 regions.

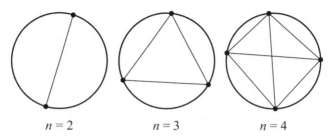

$n = 2$          $n = 3$          $n = 4$

The pattern seems clear, doesn't it? The number of regions appears to be doubling every time we add an extra point. So we suspect that with $n = 5$ there will be 16 regions.

And so there are:

$n = 5$

At which point, surely, we conclude with rather more confidence that with $n = 6$ the number of different regions will be 32.

But it isn't.
It's 31:

$n = 6$

And the general formula for the number of regions isn't the simple one we had in mind at all.

It's $\frac{1}{24}(n^4 - 6n^3 + 23n^2 - 18n + 24)$.

And that's why mathematicians need *proof*.

CHAPTER THREE

## *But . . . that's Absurd . . .*

At the end of *The Adventure of the Beryl Coronet*, Sherlock Holmes explains his methods of deduction, as usual, and remarks:

> It is an old maxim of mine that when you have excluded the impossible, whatever remains, however improbable, must be the truth.

This is, in a way, like *proof by contradiction*, which is one of the most elegant and powerful techniques in the whole of mathematics.

The basic idea is to prove that some proposition is true by exploring the possibility that it is false, and then showing that this would lead to a contradiction or nonsense of some kind. So the proposition can't be false, and the only possibility then left is that it is true.

This whole line of argument is sometimes called the 'indirect' method of proof, or *reductio ad absurdum*.

---

As a first example, consider the so-called *Königsberg Bridge Problem*, which came to the attention of the great Swiss mathematician Leonhard Euler in 1736.

At the time, Königsberg was a town in East Prussia, divided by the River Pregel into several parts which were connected by seven bridges.

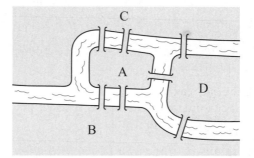

The citizens of Königsberg crossed these bridges on their long, leisurely Sunday afternoon walks. And they were vexed – so the story goes – by one particular question: can you take a walk in Königsberg in such a way that you cross each of the seven bridges *once and only once*?

Now, at first sight we are faced with the tedious and daunting task of considering all the possible routes in turn, and showing that none of them works. But, as Euler showed, there is a clever way of circumventing all this. And one convincing way of presenting the argument is as a proof by contradiction.

Suppose, then, that it *is* possible. We start, in other words, in one of the four regions A, B, C, D and end up at one of them (possibly the same one), having crossed each of the seven bridges exactly once.

Now, it follows immediately that there will be at least two regions which are neither at the beginning nor at the end of the walk. Consider one of these regions. We visit it a certain number of times and leave it an equal number of times, and as we cross each bridge exactly once it follows that there must be an even number of bridges leading from this region.

But no region in the Königsberg figure above has this property: the island A has 5 bridges leading from it, while the regions B, C and D all have 3 each.

So you *can't* take a walk in Königsberg in this particular way.

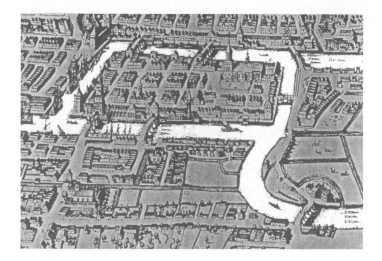

At least, you couldn't in 1736. As I understand it, the situation has now changed; for Königsberg is now Kaliningrad, and has only five bridges, most of them the result of rebuilding after the Second World War.

---

For a deeper example of proof by contradiction we turn to the subject of *prime numbers*.

Now, a prime number is a whole number, larger than 1, which is divisible only by 1 and itself. So

$$2, 3, 5, 7, 11, 13, 17, 19 \ldots$$

are all prime, but 15, for example, is not, because it is divisible by 3 and by 5.

| 2 | 3 | 5 | 7 | 11 | 13 | 17 | 19 | 23 | 29 |
|---|---|---|---|----|----|----|----|----|----|
| 31 | 37 | 41 | 43 | 47 | 53 | 59 | 61 | 67 | 71 |
| 73 | 79 | 83 | 89 | 97 | 101 | 103 | 107 | 109 | 113 |
| 127 | 131 | 137 | 139 | 149 | 151 | 157 | 163 | 167 | 173 |
| 179 | 181 | 191 | 193 | 197 | 199 | 211 | 223 | 227 | 229 |
| 233 | 239 | 241 | 251 | 257 | 263 | 269 | 271 | 277 | 281 |
| 283 | 293 | 307 | 311 | 313 | 317 | 331 | 337 | 347 | 349 |
| 353 | 359 | 367 | 373 | 379 | 383 | 389 | 397 | 401 | 409 |
| 419 | 421 | 431 | 433 | 439 | 443 | 449 | 457 | 461 | 463 |
| 467 | 479 | 487 | 491 | 499 | 503 | 509 | 521 | 523 | 541 |

The first 100 prime numbers.

Every whole number greater than 1 is either prime or can be written as a product of primes. So, for instance, 17 is prime but 18 can be written as $2 \times 3 \times 3$. In this sense, primes are the 'building blocks' out of which other whole numbers can be created by multiplication.

As we proceed up the list of whole numbers, primes occur quite frequently at first, but less frequently later on. Thus 25% of the numbers up to 100 are prime, but the corresponding figure for numbers up to 1,000,000 is just 7.9%.

An obvious question, then, is: does the list of primes come to a complete stop somewhere, or does it go on for ever?

And Euclid discovered the answer: *there are infinitely many prime numbers.*

How, then, did he *prove* it?

"YOU WANT PROOF? I'LL GIVE YOU PROOF!"

The answer is that he turned to proof by contra-diction.

He began, then, by supposing that the number of primes is finite, in which case there will be some *largest* prime number, which we will call $p$. The complete list of primes will then be

$$2, 3, 5, 7, 11, 13, \ldots, p.$$

So far, so good. Even straightforward, you might say. But the next step is an inspired one.

Euclid's ingenious idea was to consider the number

$$N = 2 \times 3 \times 5 \times \ldots \times p + 1$$

i.e. the number obtained by multiplying all the primes together and adding 1.

Now, this number is certainly greater than $p$, and as $p$ is the largest prime this new number $N$ cannot be prime. It must therefore be possible to write it as a product of primes, i.e. it must be divisible by at least one prime number.

But it isn't; if you divide $N$ by any prime number from the list $2, 3, 5, \ldots, p$ you always get a remainder of 1.

We have arrived at a contradiction, then, and the only way out is for the original hypothesis to be wrong; the number of primes cannot be finite – it must be infinite.

But some problems in number theory are more tricky.

Suppose, for instance, that we're dealing with whole numbers, and we ask whether it is possible for two square numbers to add up to a square number. After a bit of trial and error, we decide that it certainly *is* possible, for

$$3^2 + 4^2 = 5^2$$

is one example, and there are plenty of others.

But if we try to find two *cubes* that add up to a *cube*, it's quite a different matter. By trying hard enough, with some quite big numbers, we come across some amusing 'near misses'. For instance

$$729^3 + 244^3 = 401,947,273$$

while

$$738^3 = 401,947,272$$

which is 'almost' there . . . but *not quite*. And, try as we will, we just can't seem to find whole numbers $a$, $b$, $c$ so that $a^3 + b^3 = c^3$. Not only this, but the same seems to be true of $a^4 + b^4 = c^4$.

All this was anticipated in 1637, when the French mathematician Pierre Fermat scribbled in a textbook the truly sweeping claim:

> *Fermat's Last Theorem*: It is impossible to find whole numbers $a$, $b$, $c$ such that
>
> $$a^n + b^n = c^n$$
>
> when $n$ is a whole number greater than 2.

Most irritatingly of all, he then added:

> I have a truly marvellous demonstration of this proposition, which this margin is too narrow to contain.

THE GUARDIAN
Thursday June 24 1993

*For more than 300 years Fermat's Last Theorem has tantalised scholars and amateurs alike. Yesterday a shy British professor quietly delivered the answer*

# The number's up for maths' greatest riddle

**Andrew Granville and Ian Katz**

Summing up . . . Professor Wiles enters the mathematical maze

It's not every day that maths hits the headlines.

But if Fermat really did have such a 'demonstration', it has never been found. And it was not until 1993 that Andrew Wiles finally announced a proof of Fermat's Last Theorem, in what must surely have been the most highly-publicized mathematical event of the twentieth century.

And while his proof is accessible only to experts in the field, it still makes use of the general line of argument that we have just been discussing.

So, some 2000 years after Euclid used it to such good effect on prime numbers, the idea of proof by contradiction is still alive and well.

CHAPTER FOUR

# *The Trouble with Algebra*

I wonder why it is that people are always moaning about algebra.

Here, for instance, is the French novelist Stendhal writing in the early nineteenth century:

> In my view, hypocrisy was impossible in mathematics
> . . . What a shock for me to discover that nobody
> could explain to me how it happened that minus
> multiplied by minus equals plus!

And here, rather less coherently, is a schoolboy called Molesworth, in the book *Down with Skool!* by Geoffrey Willans and Ronald Searle, first published in 1953. Molesworth has a somewhat primitive view of life, and his spelling isn't up to much, yet he still makes *his* feelings about algebra quite clear in a section called HOW NOT TO APPROACH A MATHS MASTER:

> 'Sir sir please sir sir please?'
> 'Yes molesworth?'
> 'I simply haven't the fogiest about number six sir.'
> 'Indeed, molesworth?'

'It's just a jumble of letters sir i mean i kno i couldn't care less whether i get it right or not but what sort of an ass sir can hav written this book.'

(*Maths master give below of rage and tear across room with dividers. He hurl me three times round head and then out of the window.*)

Molesworth perceives algebra, then, as 'just a jumble of letters', which strikes me as rather sad. Perhaps no one ever bothered to show him a really good example of algebra at work.

Like explaining the '1089' trick, for instance.

---

The first step, if you recall, is to take a 3-digit number, reverse it, and subtract the smaller from the larger.

Suppose, then, that the larger of the two numbers has digits $a$, $b$, $c$; then its actual value is $100a + 10b + c$, and after 'reversing' and subtracting we will have $100a + 10b + c - (100c + 10b + a)$, which is the same as

$$100a + \cancel{10b} + c - 100c - \cancel{10b} - a = 99a - 99c$$
$$= 99(a - c).$$

As $a$ and $c$ are whole numbers, this shows, then, that at the end of the first part of the trick *we will always end up with a multiple of* 99.

Now, the 3-digit multiples of 99 are 198, 297, 396, 495, 594, 693, 792, 891, and we note at once how the first and third digits of each of these add up to 9. So, when we reverse any one of these numbers and *add* – which is the last part of the 'trick' – we get 9 lots of 100 from the first digits, 9 lots of 1 from the third digits, and 2 lots of 90 from the second digits, giving

$$900 + 9 + 180 = 1089.$$

So we have done a little mathematical conjuring, and a bit of algebra helped us along the way.

In contrast to geometry, algebra was in fact something of a late development, and the subject *as we know it today* only really emerged fully in the sixteenth century.

**And to a-**
**uoide the tedioufe repetition of thefe woo2des: is e-**
**qualle to: J will fette as J doe often in woo2ke bfe,a**
**paire of paralleles, o2 Gemowe lines of one lengthe,**
**thus:========,bicaufe noe.2. thynges,can be moare**
**equalle.   And now marke thefe nombers.**

$$1\,4.\textbackslash{}\underline{z}.\text{---}|\text{---}.1\,5.\textbf{9}\text{=====}71.\textbf{9}.$$

First appearance of the 'equals' sign ====, in Robert Recorde's
*The Whetstone of Witte* (1557).

It was not until 1557, for example, that the familiar 'equals' sign appeared, and the figure above shows an equation which we would now write as

$$14x + 15 = 71.$$

To solve this (i.e. to find the value of $x$) we can first subtract 15 from both sides, obtaining

$$14x = 56$$

and then divide both sides by 14 to find that

$$x = 4.$$

So far, so good, perhaps. Yet as soon as equations get more complicated, many people get a bit flustered.

Molesworth, for instance, cites

$$\frac{a \times b\,(c-d)}{d \times c\,(b-a)} = \frac{pq+rs}{xg-nbg}$$

as an example, claiming that this equation is 'enough to silence anybody', and, I must say, I rather agree. It isn't my idea of an algebraic equation at all; I have no idea what we're meant to be doing with it, and if this is truly the sort of thing that Molesworth used to see on the blackboard, no wonder he was confused.

My idea of a good algebraic equation is something more like

$$(x+a)^2 = x^2 + 2ax + a^2$$

This is of a quite different nature to $14x + 15 = 71$, because it is true for *any* two numbers $x$ and $a$. This can be proved using the rules of elementary algebra, and when both $x$ and $a$ are positive it can even be seen *geometrically* from the areas in the following diagram:

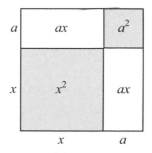

And the result is certainly useful. It allows us to solve quadratic equations, for instance, like

$$x^2 + 6x = 7.$$

By cunningly choosing $a = 3$ in our general result above, we obtain $(x + 3)^2 = x^2 + 6x + 9$, and this allows us to rewrite our quadratic equation as $(x + 3)^2 = 16$. It follows at once, then, that $x + 3$ must be either 4 or $-4$, so $x$ itself must be either 1 or $-7$.

Any quadratic equation can be solved by the same method.

---

Of course, this does rather beg the question: 'Why solve quadratic equations at all?'

Although this is a perfectly fair thing to ask, different mathematicians will, I think, answer it in different ways. As an applied mathematician who often works on the stability of mechanical systems, I can only say that I have lost count of the number of times that a problem has come down *in the end* to solving a quadratic equation.

And my mind boggles, for instance, at the number of quadratic equations that must have been solved – somewhere along the way – in the course of devising the guidance and control technology that put men on the Moon.

How many quadratic equations did it take to get to the Moon?

But if you really want to see algebra in action you can do a lot worse than combine it with geometry.

This particular development in mathematics, in the early seventeenth century, is due largely to Fermat and Descartes. They wanted to be able to convert geometric problems into algebraic ones, or vice versa. As Descartes himself put it:

> In this way, I should be borrowing all that is best in geometry and algebra, and should be correcting all the defects of the one by the help of the other.

To do this, we first draw two perpendicular *axes*, and then give each point a pair of *coordinates* $(x, y)$:

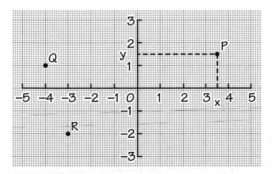

In the diagram above, for instance, the coordinates of the point P are $x = 3.5$ and $y = 1.5$, obtained by drawing lines through P which are parallel to the axes. In a similar way, the point Q has coordinates $(-4, 1)$, while those of R are $(-3, -2)$.

The main reason for doing all this is so that an equation can be represented as a curve, or vice versa.

The equation $y = 2x + 1$, for instance, corresponds to a straight line:

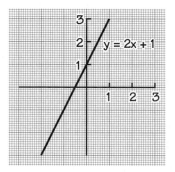

and all the points actually on the line have coordinates $(x, y)$ which satisfy the equation.

A slightly more complicated example is $y = \frac{1}{2} x^2$, which in fact corresponds to a curve called a *parabola*:

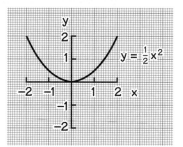

And yet another example is $x^2 + y^2 = 4$, which represents a circle:

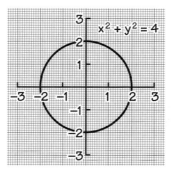

But this kind of thing wasn't the only reason that mathematicians started combining algebra and geometry in the seventeenth century.

Descartes, in particular, had another motive altogether, and once announced it in no uncertain terms:

> I have resolved to quit only abstract geometry . . . in order to study another kind . . . which has for its object the explanation of the phenomena of nature.

Which brings us nicely to the next chapter.

CHAPTER FIVE

# *The Heavens in Motion*

The appearance of unfamiliar objects in the sky has caused excitement – not to say panic – since earliest times.

Even in 1910, when Halley's comet was due, some of the American newspaper coverage struck a distinctly

alarmist note, with headlines like 'Six hours tonight in the Comet's Tail' and 'Chicago is Terrified'.

As Halley's comet has an orbital period of about 76 years, it duly made its most recent appearance in 1986. It moves very slowly when it is far away from the Sun, but speeds up greatly as it returns for a 'close encounter'.

And the shape of the orbit is an *ellipse*.

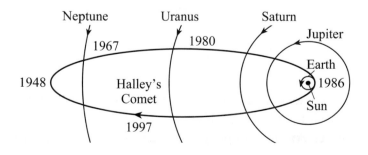

Now, the Greeks knew all about ellipses, and they knew, in particular, a simple method for constructing one. Even today, this method is occasionally used by gardeners for making flower beds.

Mark out two points H and I, and run a loop of string around them. Then keep moving the point E – as in the figure below – while keeping the string taut. The curve traced out by E will then be an ellipse.

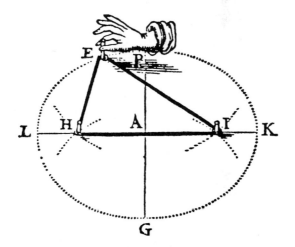

An ellipse, from van Schooten's
*Exercitationum Mathematicorum* (1657).

At first sight, perhaps, this is little more than a slightly squashed circle, but the Greeks knew that it has several interesting properties. An ellipse can be obtained, for instance, by slicing through a cone. And if we construct an elliptical mirror, and place a source of light at the point H, then any ray of light will get reflected to the point I, and vice versa. For this reason, the points H and I are known as the *focal points* of the ellipse.

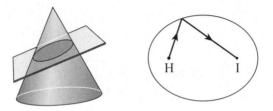

To the Greeks, then, the ellipse was a graceful curve with some interesting geometrical properties.

And so it remained, for 1500 years.

———

Then, early in the seventeenth century, the German astronomer Johannes Kepler made an extraordinary discovery.

It had been known for some time that the planets move in approximately circular orbits around the Sun, but after a painstakingly careful analysis of the observations Kepler showed that these planetary orbits are actually ellipses.

And, even, more remarkably, the Sun is always *at one of the focal points*!

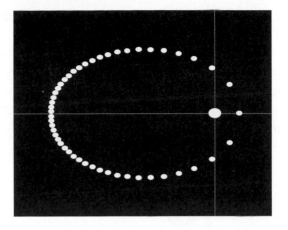

But Kepler went further still. He knew that each planet speeds up in its orbit as it gets closer to the Sun, and slows down as it moves further away. This can be seen, for example, in the picture above, which shows successive positions of the 'planet' after equal intervals of time. And he discovered a simple rule that describes precisely how this speeding up and slowing down occurs. If we imagine a line extending from the Sun to the planet, then, according to Kepler, this line rotates in such a way that it sweeps out *equal areas in equal times*.

And while it took many years for Kepler's work to become known and widely accepted, explaining all these extraordinary results eventually became the key scientific problem of the late seventeenth century.

The first big question to ask was why each planet moves in a *curved* path at all.

Now, when a stone is whirled in a sling, a force is exerted on the stone towards the centre E of the circular motion. It is this force which keeps the stone on a curved path, and without it the stone would simply fly off at a tangent, along the straight line ACG:

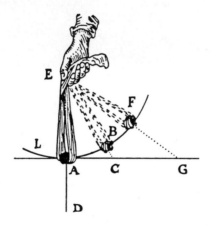

A stone being whirled in a sling (from Descartes'
*Principles of Philosophy*, 1644).

Considerations such as these led to the conclusion, then, that there must be a *force* on each planet to account for its curved orbit.

But where was this force coming from?

Gradually, it seems, the idea began to emerge that the Sun might exert a 'gravitational' force of attraction on each planet:

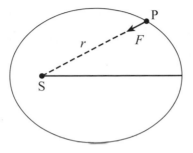

Yet, even if this were so, how would the gravitational force $F$ depend on $r$, the distance from the Sun? It seemed natural to suppose that any such force would be weaker at large distances. Some lines of evidence, based on another of Kepler's discoveries, suggested that $F$ might be proportional to $1/r^2$.

But no one really knew.

By about 1679 the whole problem was being addressed, especially in London, by some of the finest scientists of the time. These included Edmund Halley, and also Robert Hooke, who is perhaps best known for his law of elasticity and for his invention of the microscope.

Hooke kept a diary, and noted down, from time to time, progress with the planetary motion problem:

> *October 18th* Gresham College about Elliptick motion.

> *October 21st* At Bruins coffee house with Sir Chr. Wren about . . . coyled cone for Celestiall theory.

Exactly what Hooke meant by a 'coyled cone' is, I think, lost, but it is known that he had various ideas for simulating planetary motion by some kind of mechanical device.

In the end, however, it was not mechanical devices that solved the problem, but mathematics.

And the moment when things really came to a head was in August 1684, when Halley went to visit Isaac Newton.

---

Newton was Lucasian professor of mathematics at Cambridge, and while he was destined, of course, to become one of the greatest figures in the history of science, this was not entirely obvious at the time.

It was certainly not obvious to his students, who rarely attended his lectures. For, according to one of his contemporaries:

> . . . so few went to hear Him, & fewer yet understood him, that oftimes he did in a manner, for want of Hearers, read to ye Walls.

Isaac Newton (1642–1727).

Yet there is no doubt of the intensity with which he was pursuing his own scientific and mathematical research:

> He always kept close to his studyes, very rarely went a visiting, & had as few Visiters. . . . I never knew him take any Recreation or Pastime, either in Riding out to take ye Air, Walking, Bowling, or any other Exercise whatever, Thinking all Hours lost, that was not spent in his studyes. . . . He very rarely went to Dine in ye Hall . . . & then, if He has not been minded, would go very carelessly, with Shooes down at Heels, Stockins unty'd, surplice on, & his Head scarcely comb'd.

This is the kind of man, then, that Dr. Edmund Halley went to see, in 1684, about the greatest scientific problem of the day.

And the meeting with Newton was a great success, for

> ... after they had been some time together, the D$^r$ asked him what he thought the Curve would be that would be described by the Planets supposing the force of attraction towards the Sun to be reciprocal to the square of their distance from it. S$^r$ Isaac replied immediately that it would be an Ellipsis, the Doctor struck with joy & amazement asked him how he knew it, why saith he I have calculated it ...

By 1687, just three years after that meeting, Newton's 'calculation' – and many others like it – had evolved into the *Principia*, one of the most influential books on science ever written.

Quite early in the book, Newton seeks an explanation of why each planet speeds up as it gets nearer to the Sun. And he shows, with apparent ease, that Kepler's equal-area rule can be explained simply by assuming that the gravitational force on each planet is towards the Sun. Kepler's equal-area rule is, in other words, simply a consequence of the direction in which the gravitational force acts.

Determining the *magnitude* of this force, *F*, turns out to be more difficult. Eventually, however, in Proposition

XI, Newton provides the answer: if, as Kepler says, a planet moves in an ellipse, with the Sun at one focal point, then *F must be proportional to* $1/r^2$, where *r* denotes the planet's distance from the Sun.

---

### PROPOSITION XI. PROBLEM VI.

*If a body revolves in an ellipſis : it is re-
quired to find the law of the centri-
petal force tending to the focus of the
ellipſis.* Pl. 4. Fig. 2.

Let *S* be the focus of the ellipſis. Draw *SP*
cutting the diameter *D K* of the ellipſis in *E,* and the
ordinate *Q v* in *x*; and compleat the parallelogram
*Qx P R.* It is evident that *E P* is equal to the greater ſe-

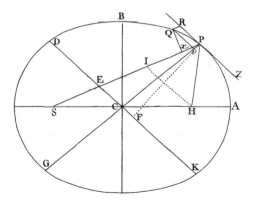

An extract from the first English edition of Newton's *Principia*, published in 1729.

The methods that Newton uses to prove this are highly geometrical; in fact, it is not overdoing it to say that any modern reader who opens the *Principia* is literally assailed by geometry on virtually every page.

At first sight, too, this geometry looks classical, just like the geometry of the ancient Greeks. Yet, on closer inspection, we see that it is slightly different; Newton is doing new things with it that had not been done before.

And while much of the *Principia* is highly technical, and full of brilliantly inventive *ad hoc* arguments, there is one underlying idea which – to modern eyes – is really quite simple.

For, time and again, Newton approaches a deep and difficult dynamical problem by imagining the whole motion as split up into a *very large number of very small changes*.

And this deceptively simple idea is the basis for one of the greatest advances in the whole of mathematics, as we see next.

CHAPTER SIX

# *All Change!*

All around us, things are changing.

Whether the 'thing' in question is the position of a tennis racket, the value of a stock market share, or the pressure in a blood vessel, change is everywhere.

And the branch of mathematics which is most concerned with change is *calculus*.

The key idea of calculus is in fact not so much change itself, but rather the *rate* at which change occurs.

Imagine, then, that we have some quantity $y$ which changes with time $t$.

Mathematicians denote the *rate* at which $y$ increases by the rather curious symbol

This symbol, pronounced 'dee−why−dee−tee', is not at all what it seems, and takes a little getting used to.

And the best way of doing this is to see how it is actually calculated.

———————

We begin by considering *small changes* in the various quantities involved.

And it is convenient to introduce a sort of shorthand for these small changes, by letting the Greek letter δ, i.e. 'delta', denote the phrase 'small change in'. Thus δ$t$ denotes a small change in time, and if, for example, $t$ were to change from the value 1 to the value 1.01, then δ$t$ would be 0.01.

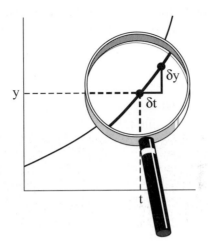

Now, after a small change in time, $\delta t$, the quantity of interest $y$ will, itself, have changed by a small amount $\delta y$, and the next step is to simply divide one change by the other to form the quantity $\frac{\delta y}{\delta t}$.

Although this is beginning to look like the quantity we want, we are not in fact quite there. And the last step is a subtle one.

For, to obtain $\frac{dy}{dt}$, i.e. the *rate* at which $y$ is changing, we take the quantity $\frac{\delta y}{\delta t}$, and find what value it approaches as both $\delta y$ and $\delta t$ become *smaller and smaller*.

And the best way of seeing how this really works is by taking a simple example.

Imagine, then, that a toy train accelerates from a standing start, so that after $t$ seconds it has travelled a distance $y$ cm, where

$$y = t^2.$$

Now, the train certainly is accelerating, because after 1 second it has travelled 1 cm, but after 2 seconds it has travelled $2^2 = 4$ cm, i.e. not twice as far but four times as far.

The speed of the train is changing all the time, then, and the question arises: what *is* the speed of the train at time $t$?

And as speed is just rate of increase of distance, this amounts to the purely mathematical question: if $y = t^2$, what is $\frac{dy}{dt}$?

Well, at time $t$ the train will have travelled a distance $y = t^2$, and a small time later, at time $t + \delta t$, it will have travelled a distance $y + \delta y = (t + \delta t)^2$. The extra distance travelled, $\delta y$, will therefore be $(t + \delta t)^2 - t^2$, which is the same as $t^2 + 2t \times \delta t + (\delta t)^2 - t^2$. It follows then that $\delta y = 2t \times \delta t + (\delta t)^2$. And on dividing this extra distance travelled, $\delta y$, by the extra time taken, $\delta t$, we find that

$$\frac{\delta y}{\delta t} = 2t + \delta t.$$

As mentioned earlier, the last step in the procedure is to consider what happens to this quantity $\frac{\delta y}{\delta t}$ as the small change in time, $\delta t$, is taken smaller and smaller. And, plainly, as we do this, $\frac{\delta y}{\delta t}$ gets closer and closer to the value $2t$.

In purely mathematical terms, then, we have shown that

$$\text{If } y = t^2, \text{ then } \frac{dy}{dt} = 2t.$$

And from a more 'practical' view, we have calculated the speed of the train; at time $t$ the speed is $2t$.

G.W. Leibniz (1646–1716), who, together with Newton and
others, discovered the calculus, as we know it today.


Similar methods can be used to calculate the rates at
which other quantities $y$ change with time. The whole
process is called *differentiation*, and a few examples are
listed below.

| $y$ | $\dfrac{dy}{dt}$ |
| --- | --- |
| 1 | 0 |
| $t$ | 1 |
| $t^2$ | $2t$ |
| $t^3$ | $3t^2$ |
| $t^4$ | $4t^3$ |
| $t^5$ | $5t^4$ |

The results here appear to be following a simple pattern, and it is in fact true that if $y = t^6$ then $\frac{dy}{dt} = 6t^5$, and so on. And we shall use this particular property, in a spectacular way, at the end of the book.

---

For the time being, however, it is the broad *ideas* of calculus that really matter, which, as I see it, have just been presented in a concise and uncompromising way.

To recap, mathematicians denote the *rate* at which some quantity $y$ changes with time by the symbol $\frac{dy}{dt}$.

In effect, $\frac{d}{dt}$ is itself a symbol, meaning 'rate of change of'.

Thus, whenever $y$ increases slowly with time, $\frac{dy}{dt}$ will be small, and whenever $y$ increases rapidly with time, $\frac{dy}{dt}$ will be large. And if $y$ actually decreases with time, then $\frac{dy}{dt}$ will be negative.

While we can view the rate at which $y$ changes, loosely, in these terms:

$$\frac{dy}{dt} \approx \frac{\text{small increase in } y}{\text{small increase in } t},$$

the real truth, as we have seen, is rather more subtle.

And after their first appearance, in the seventeenth century, all these ideas were to gradually open up so many new lines of enquiry that neither mathematics nor physics would ever be quite the same again.

# CHAPTER SEVEN

# *On Being as Small as Possible*

Why not try this simple experiment?

Take a wire framework of some kind, and dip it into a bowl of soapy water (or washing-up liquid). When you take it out again, you will find a thin soap film spanning the framework.

And however complicated this film may be, it has one very interesting property: it always tries to organize itself so that its surface area is *as small as possible*.

Now, in mathematics, problems which begin '*Find the smallest …*' (or indeed largest) often have a special appeal, not least because they can sometimes have very elegant and satisfying solutions.

Here's a simple example. Imagine, if you will, that a cowboy is returning home after a long day out on the range, and suddenly decides to take his horse to the river for a final drink.

The question is: how should he do this *so that the total distance home is as small as possible*? In other words, which point of the riverbank should he pick in order to minimize the total distance home?

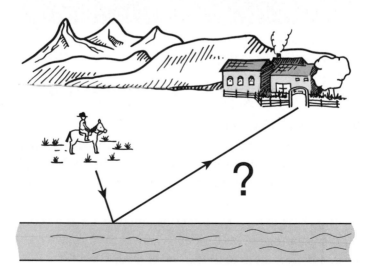

The answer, in fact, is that he should choose his 'outward' and 'return' paths in such a way that they *make equal angles with the river*.

And to see this, the trick is to imagine that the homestead H is at the same distance from the riverbank, *but on the opposite side*, at H'. For, whichever point P on the riverbank the cowboy stops at, the distances PH and PH' will then be equal. So the problem of picking P so as to minimize CP + PH is the same as the problem of picking P so as to minimize CP + PH'.

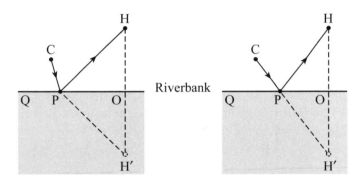

But this 'new' problem is easy; we solve it by picking P so that CPH' is a straight line. And in that case the angles OPH' and QPC will be equal. As the angles OPH' and OPH will be equal *anyway*, no matter where P is on the riverbank, it follows that the shortest path CP + PH will be achieved when angle QPC = angle OPH.

Clever arguments like this are all very well, but mathematicians need general methods, too, for tackling maximum or minimum problems. And one of the best-known of these involves *calculus*.

To see how this works, imagine that a farmer has 4 km of fencing, and wants to construct a rectangular field in such a way that the area of the field is as large as possible.

Now, if two of the sides are of length $x$, say, then the other two must be of length $2 - x$, so the area will be $x(2 - x)$, which can be rewritten as $2x - x^2$.

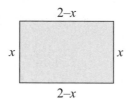

The farmer's problem amounts, then, to finding the value of $x$ which makes the quantity

$$y = 2x - x^2$$

as large as possible.

And this is where calculus comes in, because a change in $x$ will produce a change in $y$, and we can use the methods of Chapter 6 to deduce the *rate* at which $y$ changes with $x$:

$$\frac{dy}{dx} = 2 - 2x.$$

So, if $x$ is less than 1, $\frac{dy}{dx}$ is positive and $y$ increases with $x$, but if $x$ is greater than 1 then $\frac{dy}{dx}$ is negative and $y$ decreases with $x$. Not only does this help us sketch the graph of $y$ against $x$:

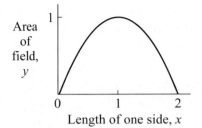

Length of one side, $x$

but it tells us, of course, that the maximum value of $y$ must occur when

$$\frac{dy}{dx} = 0,$$

i.e. when $x = 1$, because that is the point at which $y$ stops increasing with $x$ and starts decreasing.

To put the matter a slightly different way, $\frac{dy}{dx}$ is called the *slope* of the curve at any point – a measure of its steepness – and it is at $x = 1$ that this slope changes from being positive to being negative.

And if we recall that the farmer's field has sides of length $x$ and $2 - x$, we see that $x = 1$ corresponds to a square field.

So, a square field is 'best'.

In both of the problems we have considered so far, the task has effectively been to find one number – the distance of the point P along the riverbank in the first problem, and the value of $x$ in the second.

But sometimes in mathematics the task is to find a whole *curve*, or *surface*, even, such that some quantity is a maximum or a minimum.

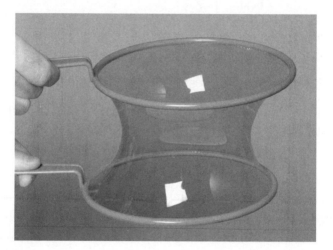

Suppose, for instance, we take two circular hoops, dip them into soapy water and then take them out again. A thin soap film will span the gap between the hoops, and, as mentioned earlier, it will try to arrange itself so that its surface area is as small as possible. Given, then, the radius of the hoops, and their distance apart, how can we use mathematics to determine the shape of the soap film which has this minimum-surface-area property?

This is a seriously difficult problem, best tackled using a quite sophisticated branch of mathematics called the *calculus of variations*.

And the same is true of another famous problem, posed by the Swiss mathematician Johann Bernoulli in 1696.

In this case, a bead slides down a wire between two given points A and B:

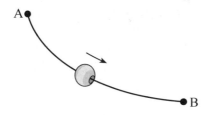

It is not 'pushed' at all; it is initially stationary, and just slides down under its own weight. We assume, too, that there is no friction.

And the question is: which curve allows the bead to slide from A to B *in the shortest possible time*?

(It is tempting, perhaps, to guess that the answer will be the straight line between A and B, but while this is certainly the path of shortest *distance* it is not in fact the path of shortest time.)

Bernoulli knew the solution, and challenged his mathematical contemporaries to solve the problem. It went down very well with the Marquis de l'Hospital, for instance, who wrote back at once saying:

> This problem seems to be one of the most curious and beautiful that has ever been proposed, and I would very much like to apply my efforts to it, but for this it would be necessary that you reduce it to pure mathematics, since physics bothers me...

Isaac Newton, on the other hand, rose to the challenge with rather less good temper, and was apparently heard to mutter that he did not like to be

> ... teezed by forreigners about Mathematical things.

The solution to the problem is in fact a *cycloid*. This is the curve traced out by a point on the rim of a wheel which rolls on a flat surface:

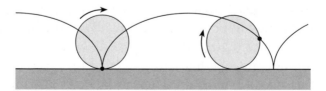

The idea, then, is to construct a cycloid of the right size, turn it upside-down, and fit it through the given points A and B:

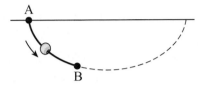

And the most curious feature of all is that the relative positions of A and B may be such that the quickest-descent curve first passes *below* B, before turning up again:

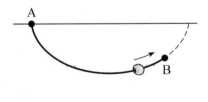

We end this chapter, though, with a problem which *sounds* innocent enough, but is in fact so tricky that, in its most general form, it can defeat even the fastest of modern computers.

The problem is: how do you connect a number of different towns by a road network which is as short as possible?

To get some idea of where the difficulty lies, consider the 'simple' case in which there are just four towns A, B, C, D lying – somewhat conveniently – at the corners of a square of side 1.

Now, our only requirement is that people from any one town should be able to travel to any other town. So we could just build a road network consisting of 3 straight lines:

But this network, of length 3 units, certainly isn't the shortest network connecting the four towns. After a little trial and error, we soon find that we can do better by introducing an intersection point in the middle, and making use of the diagonals:

This is because AC and BD will both be equal to $\sqrt{2}$, by Pythagoras' theorem, so the total length of this new network will be $2\sqrt{2} = 2.83$.

And of course this immediately raises the question of whether we might do better still with more than one intersection point. But, if so, then how many, exactly, and where should we put them?

These are seriously difficult questions, and one way of dealing with them is to *cheat*, and use soap films.

Suppose, then, that we take two parallel Perspex plates and join them by four pins at the corners of a square. Each time we immerse this apparatus in a bowl of soapy water, and take it out again, we get a soap film with a surface area that is smaller than it would be in any slightly different state. Each time, in other words, we get a serious *candidate* for the solution to our road-network problem.

And, sooner or later, a particularly distinctive soap film configuration emerges:

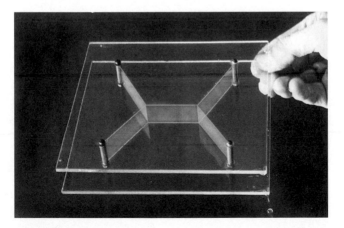

And while a purely mathematical proof isn't easy, this is in fact the solution to our road network problem, i.e. five straight portions and two 3-way intersections at angles of 120°:

The total length of the network is then $1 + \sqrt{3} = 2.73$ units, and no shorter connecting network is possible.

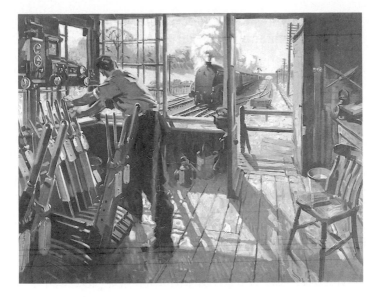

# CHAPTER EIGHT

# *'Are We Nearly There?'*

Processes which go on *for ever* are quite common in mathematics, and the subject as a whole would be very different without them.

Consider, for instance, the 'sum'

$$\tfrac{1}{2} + \tfrac{1}{4} + \tfrac{1}{8} + \tfrac{1}{16} + \dots$$

where the dots indicate that we are to keep on adding extra terms, in the obvious way, *without ever stopping*.

At first sight, perhaps, this 'sum' is bound to be infinite, because all the individual terms are positive numbers, and we are adding up an infinite number of them.

But imagine, for instance, dividing up a cake by first taking half the cake, then a quarter, then an eighth, and so on:

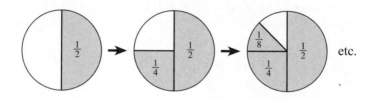

It is very noticeable that with each new piece we take, in this particular way, the amount of cake left over is halved.

Suddenly, then, two things seem clear. First, we are never going to get the whole cake by this procedure. Second, we can, however, get *as much of it as we like by taking enough pieces.*

And this is, essentially, exactly what mathematicians mean when they say that

$$\tfrac{1}{2} + \tfrac{1}{4} + \tfrac{1}{8} + \tfrac{1}{16} + \dots$$

*converges* to the value 1.

In this sense, then, it is certainly possible for an infinite series of terms to have a finite 'sum'.

But it won't always happen, and a famous cautionary example of this is the series

$$\frac{1}{2} + \frac{1}{3} + \frac{1}{4} + \frac{1}{5} + \frac{1}{6} + \ldots$$

Once again, each term is smaller than the one before, but this time the terms are not getting smaller *fast enough*, and the series does not converge to a finite sum.

And there is a very simple, elegant argument which shows this. All we have to do is group terms of the series together in the following way:

$$\frac{1}{2}$$
$$+ \frac{1}{3} + \frac{1}{4}$$
$$+ \frac{1}{5} + \frac{1}{6} + \frac{1}{7} + \frac{1}{8}$$
$$+ \frac{1}{9} + \frac{1}{10} + \frac{1}{11} + \frac{1}{12} + \frac{1}{13} + \frac{1}{14} + \frac{1}{15} + \frac{1}{16}$$
$$\vdots$$

so that each new group has twice as many terms as the previous one. We then observe that $\frac{1}{3} + \frac{1}{4}$ is greater than $\frac{1}{4} + \frac{1}{4} = \frac{1}{2}$, that the next group is greater than $\frac{1}{8} + \frac{1}{8} + \frac{1}{8} + \frac{1}{8} = \frac{1}{2}$, that the one after that is greater than $8 \times \frac{1}{16} = \frac{1}{2}$, and so on. And as $\frac{1}{2} + \frac{1}{2} + \frac{1}{2} + \ldots$ doesn't converge to a finite sum, it follows that the series in question can't, either.

So, while infinite series can be intriguing, they do need to be handled with care.

*Great* care.

---

Another quite different way in which infinite processes can arise is through the whole problem of *area*.

Imagine, for instance, that we want to find the area of a region with a curved boundary. It is not entirely clear what this concept means, let alone how we should go about calculating it.

On the other hand, we do know how to calculate the area of a *rectangle* easily enough, so one way forward would be to fill the region in question with lots of thin rectangles.

And, fairly evidently, the more of these we have, and the thinner they are, the better.

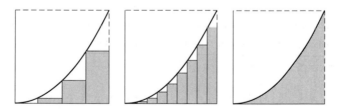

In this way, we run once again into a mathematical process that, in a sense, goes on for ever.

---

But infinite processes in mathematics aren't just about summing series or finding areas. They arise, in a sense, as soon as we start thinking seriously enough about the nature of *number* itself.

For mathematicians like to imagine their numbers sitting on a continuous *number line* with each and every number in its proper place:

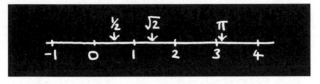

First come the whole numbers, or *integers*: 0, ±1, ±2, ... We then generate more numbers, such as $\frac{7}{2}$, by dividing each of these units into two equal parts. We then generate still more of these *fractions* by dividing each unit into three equal parts ... and so on.

And we might well imagine that by dividing up each unit into a larger and larger number of smaller and smaller *equal* parts we can reach eventually *all* the numbers on the line in this way.

But we can't.

Even though it goes on 'forever', the procedure will only pick up all the so-called *rational* numbers, i.e. numbers which can be written as a ratio of two integers. It turns out that there are other numbers – called *irrational* numbers – that cannot be written in this way.

And $\sqrt{2}$ is one of them.

# makes things **whiter!**

The argument is in fact a classic example of proof by contradiction. We begin, in other words, by supposing that $\sqrt{2}$ *can* be written as a fraction. Then, by reducing that fraction to its 'lowest terms', i.e. by cancelling out any common factors, we obtain $\sqrt{2} = m/n$, where $m$ and $n$ are whole numbers which have no common factor.

To see a contradiction develop, begin by squaring both sides to obtain $2 = m^2/n^2$, so that $m^2 = 2n^2$. This means that $m^2$ is twice a whole number, so $m^2$ is even. It follows that $m$ *must be even* (for if $m$ were odd, $m^2$ would be odd, as odd $\times$ odd = odd).

Now , as $m$ is even, it can be written as $2r$, where $r$ is a whole number. The equation $m^2 = 2n^2$ can then be rewritten as $4r^2 = 2n^2$, i.e. $n^2 = 2r^2$. So $n^2$ is even, and by the same argument as before, *n must be even*.

And there is the contradiction: $m$ and $n$ started by having no common factor, yet must now have a common factor of 2, because they are both even.

The only way out of this absurd situation is for the original assumption – that $\sqrt{2}$ can be written as a ratio of two whole numbers – to be false.

So $\sqrt{2}$ is an irrational number. And there are plenty of others; there is nothing exceptional or peculiar about them. In fact, there are 'more' irrational numbers than there are rational ones, though what this statement means, exactly, takes a bit of thinking about, as the two things we are comparing are both infinite.

---

Sometimes in mathematics, an infinite process can be involved in the actual structure of the logical reasoning itself. This happens, for example, with a powerful method called *proof by induction*.

The general idea of this is not unlike a railway train; a lot of coaches are coupled together, an engine pulls on the first one, this coach then pulls on the second, and so on until the whole train moves.

Here's an example. There is a simple formula for the sum of the first $n$ whole numbers:

$$1 + 2 + 3 + 4 + \ldots + n = \tfrac{1}{2}n(n+1).$$

According to this formula, then, the sum of the first 10 whole numbers is $\tfrac{1}{2} \times 10 \times 11 = 55$, and it is easy to check by direct summation that this is correct. But how can we prove that the formula is correct for *any* whole number $n$?

Well, suppose for a moment that we knew it to be true for some *particular* whole number $n = p$. If that were so, we could then deduce, simply by adding one more term, that the sum of the first $p + 1$ whole numbers must be:

$$1 + 2 + 3 + 4 + \ldots + p + (p+1) = \tfrac{1}{2}p(p+1) + (p+1).$$

And there is something very interesting about the right-hand side of this equation: with a little algebra we can rewrite it in the form $\tfrac{1}{2}(p+1)(p+2)$.

But this is just the original formula $\tfrac{1}{2}n(n+1)$ *with* $n = p+1$ *instead of* $n = p$.

We have shown, in other words, that *if* the formula happened to be true for one particular whole number $n$, *then it would be true for the next one as well.*

At this stage we have, so to speak, coupled all the coaches, and the final step is to start the engine.

And to do this, we simply observe that the formula certainly works when $n = 1$, because the 'sum' then has only one term, 1, and $\tfrac{1}{2}n(n+1) = \tfrac{1}{2} \times 1 \times 2$, which is,

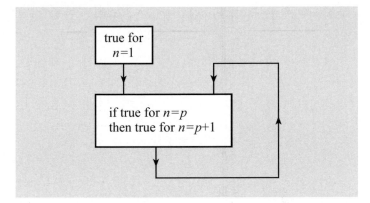

true for
$n=1$

if true for $n=p$
then true for $n=p+1$

The idea of proof by induction.

indeed, 1. From what we have just shown, then, the formula must also be true for $n = 2$ as well, and, in the same way, because it is true when $n = 2$ it must also be true when $n = 3$, and so on.

So the sum of the first $n$ whole numbers is $\frac{1}{2}n(n+1)$, for all positive whole numbers $n$.

And while there are other, equally attractive, ways of proving this particular result, the whole idea of proof by induction is a very general one, and finds application, from time to time, in countless different branches of mathematics, even at the highest level.

# CHAPTER NINE

# A Brief History of π

When we first meet the number $\pi = 3.14159\ldots$ it is all about circles. In particular, if we have a circle of radius $r$, then

$$\text{circumference} = 2\pi r,$$

and

$$\text{area} = \pi r^2.$$

The first of these formulae is more or less what we *mean* by the number $\pi$. For if we regard it as 'obvious' that the circumference of a circle is proportional to its diameter, then the ratio $\frac{\text{circumference}}{\text{diameter}}$ will be a single number, the same for all circles. And that number is denoted by the symbol $\pi$. To put it another way, we *define* $\pi$ to be that number, and as the diameter of a circle is twice the radius, i.e. $2r$, the formula *circumference* = $2\pi r$ then follows immediately.

But the second formula, *area* = $\pi r^2$, is quite a different matter. There was no mention of area at all in our definition of $\pi$ just now. Here, then, we have a simple but far from obvious result.

So why is it true?

Begin by inscribing within the circle a polygon with $N$ equal sides.

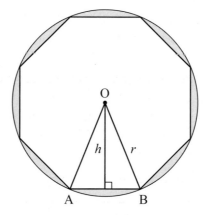

Now, this polygon will consist of $N$ triangles such as OAB, where O is the centre of the circle, and the area of each such triangle will be $\frac{1}{2}$ its 'base' AB times its 'height' $h$. The total area of the polygon will be $N$ times this, i.e. $\frac{1}{2} \times$ (AB) $\times h \times N$. But (AB) $\times N$ is the length of the perimeter of the polygon, so

$$Area\ of\ polygon = \frac{1}{2} \times (Perimeter\ of\ polygon) \times h.$$

Consider, finally, what happens as we let $N$ get larger and larger, so that the polygon has an ever-increasing number of shorter and shorter sides, and therefore approximates the circle ever more closely:

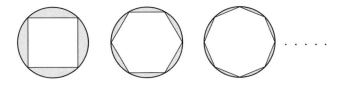

As we continue in this way, the perimeter of the polygon will get ever closer to the circumference of the circle, which is $2\pi r$, and $h$ will get ever closer to the radius of the circle, $r$. The area of the polygon will therefore get ever closer to $\frac{1}{2} \times 2\pi r \times r$.

And that is why the area of a circle is $\pi r^2$.

There are, of course, plenty of practical applications of $\pi$.

Imagine, for instance, that we have a cylindrical soup can of radius $r$ and height $H$. Not surprisingly, $\pi$ comes in to the formulae for the volume and surface area:

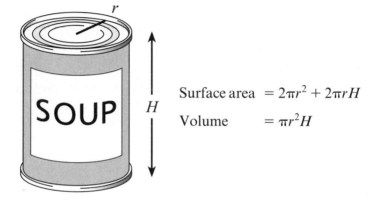

Surface area $= 2\pi r^2 + 2\pi r H$

Volume $\quad\quad= \pi r^2 H$

And an obvious question of economy arises: how do we make a can of given volume *using as little material as possible*? In particular, in order to minimize the surface area, should the can be tall and thin, or should it perhaps be short and stubby?

This problem can be tackled by the calculus method described in Chapter 7, and it turns out that the way to minimize the surface area, for a given volume, is to choose $2r = H$, so that the diameter of the can is equal to its height.

Interestingly, the tins of sweetcorn in my kitchen cupboard have precisely this shape, but the cans of soup do not.

I haven't yet discovered the reason for this, and, in any case, $\pi$ isn't really about soup. In general, it isn't even really about circles.

The truth is, $\pi$ has a habit of popping up all over mathematics, even when there isn't a circle in sight.

And to see what $\pi$ really is about, it's no bad idea to take a look at the many attempts, throughout history, to establish its precise numerical value.

---

The earliest known estimate of $\pi$ is $(\frac{4}{3})^4 = 3.16 \ldots$, which appears in the Rhind Papyrus, dating from about 1650 BC. Despite this, the crude approximation $\pi = 3$ was used throughout much of the ancient world, and this is the approximation which appears in the Old Testament:

> . . . . Also he made a molten sea of ten cubits from brim to brim, round in compass . . . and a line of thirty cubits did compass it round about.
>
> (1 Kings 7: 23)

The first really systematic attempt to pin down the value of $\pi$ seems to have been by Archimedes, who used polygons with 96 sides, inside *and* outside the circle, to show that $\pi$ must be greater than $3\frac{10}{71}$ but less than $3\frac{1}{7}$. And this upper bound of $\frac{22}{7}$ often appeared, centuries later, as an approximation to $\pi$ in elementary textbooks.

The first exact formula for $\pi$ was obtained in 1593 by Viète:

$$\frac{2}{\pi} = \frac{\sqrt{2}}{2} \times \frac{\sqrt{2+\sqrt{2}}}{2} \times \frac{\sqrt{2+\sqrt{2+\sqrt{2}}}}{2} \cdots$$

and this remarkable infinite product was again derived by considering polygons. The square roots make it a little cumbersome, but still permitted, even in Viète's time, the numerical calculation of $\pi$ to 14 decimal places:

$$\pi = 3.14159\ 26535\ 8979 \ldots$$

---

The whole approach changed completely with the advent of *calculus* in the mid-seventeenth century, and one of the first formulae for $\pi$ to emerge from the new methods was another infinite product:

$$\frac{\pi}{2} = \frac{2}{1} \times \frac{2}{3} \times \frac{4}{3} \times \frac{4}{5} \times \frac{6}{5} \times \frac{6}{7} \cdots$$

obtained by John Wallis in 1655. Unlike Viète's result, this contains no square roots, and it is also much more obvious that successive factors are getting closer and closer to 1, which is how the infinite product manages to converge to a finite value.

A little later, in 1674, Leibniz published the famous infinite series

$$\frac{\pi}{4} = 1 - \frac{1}{3} + \frac{1}{5} - \frac{1}{7} + \cdots$$

linking π with the odd numbers, though it is now known that Keralese Indian mathematicians had discovered this – in a quite different way – over 150 years earlier.

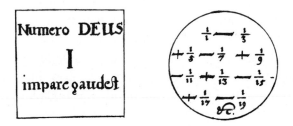

Illustrations from Leibniz's 1674 paper. Loose translation: 'God loves odd numbers'.

While the series has a breathtaking simplicity, it is pretty useless as a practical method for calculating $\pi$, because it converges so slowly. Even after 300 terms, for instance, we have an estimate for $\pi$ which is less accurate than Archimedes' approximation of $\frac{22}{7}$, obtained 2000 years earlier!

Another famous infinite series in which $\pi$ makes a completely unexpected appearance is

$$1 + \frac{1}{2^2} + \frac{1}{3^2} + \frac{1}{4^2} + \frac{1}{5^2} + \ldots = \frac{\pi^2}{6}$$

which Euler obtained by a superbly reckless argument in 1736.

———————

By Euler's time, $\pi$ had been determined to about 100 decimal places by series methods, but in 1761 Lambert finally proved what everyone had long suspected: $\pi$ is *irrational*, so cannot be expressed exactly as the ratio of two whole numbers. This implies, in particular, that the decimal expansion of $\pi$ can never terminate. Notwithstanding this, modern computers have now allowed the determination of $\pi$ to several billion decimal places.

If, on the other hand, you are content with just one or two places of decimals, then you might find a *probability* approach to $\pi$ a bit simpler and more entertaining.

Leonhard Euler (1707–1783)

Take a sheet of paper ruled with straight lines a
distance $d$ apart, and drop on to it a straight pin, also of
length $d$. Then the probability that the pin will lie across
one of the lines is $2/\pi$.

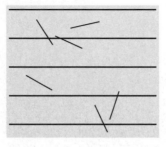

No pins to hand? Then you could try tossing a coin a
few times. (Well, rather a lot of times, actually.) If you
toss a coin $2n$ times, where $n$ is very large, then the probability of getting exactly $n$ heads and exactly $n$ tails is
approximately $\frac{1}{\sqrt{n\pi}}$.

No coins either? Then you could just ask two friends
to choose a lot of whole numbers; the probability that
any two positive integers, chosen at random, have no
common factor (other than 1) is $6/\pi^2$.

It all seems a long way from

$$\pi = \frac{\text{circumference}}{\text{diameter}}$$

# CHAPTER TEN

# *Good Vibrations*

I play a bit of jazz guitar in my spare time, and my great hero is Django Reinhardt.

Django was a Belgian gypsy, and made his name as a jazz guitar genius in the late 1930s with the Quintette du Hot Club de France, pictured above. (Django is second from the right.) Despite losing the use of two fingers in a caravan fire, he had a devastating guitar technique, and was famous for blisteringly fast solos. But he was capable of wonderfully slow, lyrical playing as well, and one of my favourites is a version of 'Body and Soul' that he recorded in Paris in 1938 with the great mouth organ player Larry Adler.

I realize, of course, that this is all rather incidental, yet I know, too, that many readers will share my love of music in one way or another. And the truth is: music is essentially made up of *vibrations*.

And as soon as we begin to think about these carefully enough, one particular type of vibration pops up everywhere.

It is called a *sine wave*:

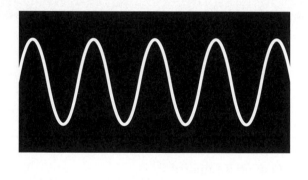

Now, the whole idea of *sine* first arose in geometry.

Suppose we have a right-angled triangle, and that one of the other angles is A, say. Then the sine of A (written sin A) is the length of the opposite side divided by the length of the longest side, or hypotenuse.

In a similar way, the cosine of A is the adjacent side divided by the hypotenuse.

$$\sin A = \frac{\text{opposite}}{\text{hypotenuse}}$$

$$\cos A = \frac{\text{adjacent}}{\text{hypotenuse}}$$

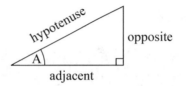

Note that the actual size of the triangle doesn't matter here; the quantities sin A and cos A depend only on the angle A, and the way in which they do this can be displayed in graphical form:

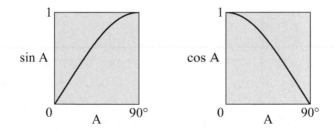

More often than not, though, when mathematicians write things like sin θ and cos θ, they're not really thinking of θ as an angle. In fact, they're not really thinking of geometry at all.

Instead, θ is viewed simply as a number, which can be as large as we like, and the quantities sin θ and cos θ are then thought of as defined by the following *graphs*:

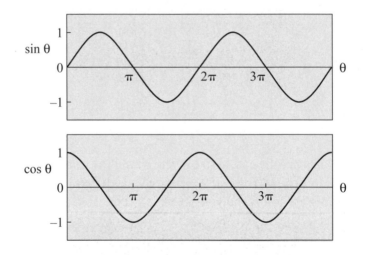

An important thing about these curves is that they begin in exactly the same way as those on p. 95; they begin, in other words, by having exactly the same shape. But they continue thereafter in an oscillatory way, and the scale is different, too: the number θ only has to increase from 0 to π/2 for sin θ to increase from 0 to 1.

One link between $\sin\theta$ and $\cos\theta$ is immediately evident, for each graph can be obtained from the other simply by shifting it along by an amount $\pi/2$. But an even more remarkable relationship between the two quantities emerges as soon as we consider the rate at which they change with $\theta$.

For it turns out that

$$\frac{d}{d\theta}\left(\sin\theta\right) = \cos\theta$$

In other words, the rate at which $\sin\theta$ increases with $\theta$ is $\cos\theta$! And it is 'almost' true the other way round, too, but not quite:

$$\frac{d}{d\theta}\left(\cos\theta\right) = -\sin\theta$$

We can see evidence of this (as opposed to a proof) in the graphs themselves. For whenever $\cos\theta$ is positive (at $\theta = 0$, for example) $\sin\theta$ is indeed increasing with $\theta$. And the negative sign in the second equation above is just right, too, for whenever $\sin\theta$ is positive (at $\theta = \pi/2$, say) then $\cos\theta$ is decreasing as $\theta$ increases.

These two results are, arguably, the deepest and most far-reaching results linking sin θ and cos θ, and we will use them in a quite spectacular way at the very end of the book.

----

For the time being, however, the pressing question is: what has all this got to do with *vibrations*?

To answer this, I turn to a rather old and battered children's toy in my possession – a 'spider' which hangs from the ceiling by a long spring:

If I displace the spider upward by a few centimetres and let it go, it drops down below its original position, then bounces back up again, and keeps on doing this until friction with the surrounding air gradually stops the oscillation.

But the real point is this: if I plot the displacement of the spider against time, and ignore the slight effects of friction, the curve I get is precisely a cosine curve:

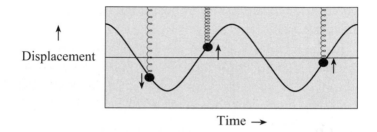

Time →

And, despite appearances, there is nothing 'peculiar' about my toy spider, either. Countless other physical systems, when disturbed slightly from their stationary or 'equilibrium' state, vibrate in exactly the same manner, with the displacement following a sine or cosine curve when it is plotted as a graph against time.

Which brings us back to guitar strings.

---

Now, if you pluck a guitar string, it typically vibrates in a very complicated way.

Yet, despite this, there are certain special 'modes of vibration' in which the oscillations are very simple, with all the different parts of the string vibrating to and fro 'in step', so to speak, and at the same, single frequency.

And whenever this happens, if we plot the displacement of any particular part of the string against time, we get a sine or cosine curve.

More remarkably still, if we take a snapshot of the whole string at any given moment, the actual *shape* of the string itself even *looks* exactly like a sine curve!

In the simplest of these modes of vibration – the so-called 'fundamental mode' – all parts of the string vibrate in the same direction at any given moment, and the largest displacement is in the middle:

The next mode – the so-called 2nd harmonic – has a frequency which is twice the fundamental, and sounds an octave higher. In this mode, one half of the string moves one way and the other half the other way, at any given moment. And there is a *node* – or point of permanently zero displacement – at the mid-point of the string:

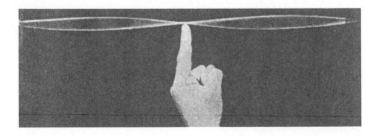

In a similar way, the 3rd harmonic vibrates at 3 times the frequency of the fundamental, and has *two* 'nodes':

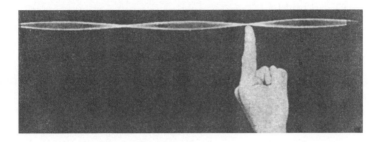

The photographs above are in fact of a vibrating rubber cord, but it is possible to generate the various harmonics on a guitar string, in essentially the way indicated, by creating a suitable node.

The trick is to lightly – and briefly – touch the guitar string in the appropriate place at just the moment that you pluck it somewhere else. For the 2nd harmonic, or 'octave', the appropriate position is half-way along the string, i.e. immediately over the 12th fret. In a similar way, touching the string just over the 7th or 19th frets can generate the 3rd harmonic.

The legendary jazz guitarist Tal Farlow made something of a speciality of playing in harmonics, and could play whole tunes, at speed, in this way. Many of us will never have the technique or, perhaps, the inclination to follow his example, but throwing in an occasional harmonic can certainly add a bit of spice from time to time, and, if I may, I would encourage guitar-playing readers to try it, if they have never done so before.

Good luck!

# CHAPTER ELEVEN

# *Great Mistakes*

Generally speaking, mathematicians are a cautious lot.

There is a story, for example, about an astronomer, a physicist and a mathematician who were on a train journey together in Scotland. Glancing from the window, they observed a black sheep in the middle of a field.

'How interesting!' said the astronomer. 'All Scottish sheep are black!'

The physicist, rather startled, said: 'Surely you mean *some* Scottish sheep are black?'

But the mathematician viewed even this as a bit rash.

'I think what you *both* mean,' he said, 'is that there is at least one sheep in Scotland which is black *on at least one side*.'

———

There is a serious point to this story, namely that, in mathematics, it is all too easy to jump to the wrong conclusion.

A good example of this is *Malfatti's problem*, where the question is: given a triangle, how do you construct three non-overlapping circles inside it so that their total area is as large as possible?

This is, in other words, a 'packing' problem, and when he first posed it, in 1803, Malfatti thought he knew the answer: you choose the circles in such a way that each one touches two sides of the triangle and both the other circles:

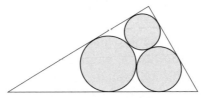

And for over a hundred years the problem was considered solved. It was hardly the most pressing problem in the subject, but it passed through a number of quite distinguished hands and everyone seemed reasonably happy with it.

Then, in 1930, somebody noticed something very strange: in the particular case of an equilateral triangle, Malfatti's 'solution' isn't correct. In his configuration:

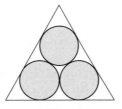

the circles occupy a fraction

$$\frac{\pi \sqrt{3}}{(1 + \sqrt{3})^2} \approx 0.729$$

of the triangle's area, but you can do slightly better by using the biggest possible circle and two smaller ones:

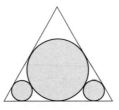

because the fraction then turns out to be

$$\frac{11 \, \pi}{27 \sqrt{3}} \approx 0.739.$$

And 35 years after that, in 1965, Howard Eves noticed something stranger still: if the triangle in question is long and thin, Malfatti's solution

is rather obviously not correct. It seems very clear, without any calculation at all, that we will do a lot better by choosing the circles as follows:

Finally, in 1967, Michael Goldberg demonstrated that Malfatti's 'solution' is *never* correct, whatever the shape of the triangle. The correct solution always has one of the forms below, with one of the circles touching all three sides:

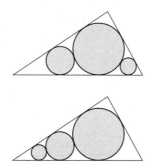

But even great mathematicians can get things badly wrong.

In 1753 Euler proved that there are no whole numbers $a$, $b$ and $c$ such that

$$a^3 + b^3 = c^3$$

To put the matter another way, if we are dealing with whole numbers, then it is impossible for two cubes to add up to a cube. This is a special case of Fermat's Last Theorem (p. 26) and – as you might say – so far, so good.

But a few years later, Euler went on to conjecture that, in a similar way, it would be impossible for three 4th powers to add up to a 4th power, or for four 5th powers to add up to a 5th power, for five 6th powers to add up to a 6th power, and so on.

Now, the actual numbers involved in checking out propositions like this get very big, very quickly, and for many years no one could prove Euler's conjecture, yet no one could disprove it, either.

And then, in 1966, some two hundred years after Euler had originally made the conjecture, L.J. Lander and T.R. Parkin finally found a counter-example, four 5th powers that add up to a 5th power:

$$27^5 + 84^5 + 110^5 + 133^5 = 144^5.$$

In mathematics, then, a wrong conjecture can *remain* a wrong conjecture for a very long time.

One area of mathematics which provides umpteen opportunities for subtle mistakes is the whole subject of *infinite series*.

Consider, for instance,

$$1 - \tfrac{1}{2} + \tfrac{1}{3} - \tfrac{1}{4} + \tfrac{1}{5} - \tfrac{1}{6} + \ldots$$

This series does, in fact, converge, and its sum turns out to be 0.693. . . .

But suppose that we now add up the terms of the series *in a different order*, by having each positive term followed by two negative ones:

$$\left(1 - \tfrac{1}{2}\right) - \tfrac{1}{4} + \left(\tfrac{1}{3} - \tfrac{1}{6}\right) - \tfrac{1}{8} + \left(\tfrac{1}{5} - \tfrac{1}{10}\right) - \tfrac{1}{12} + \ldots$$

We stress that all the terms of the series are 'still there'; we haven't missed out any, and we haven't smuggled in any new ones, either.

It would seem, then, that the new series ought to have the same sum as before.

But it doesn't. By simplifying the terms in brackets we can rewrite the new series as $\tfrac{1}{2} - \tfrac{1}{4} + \tfrac{1}{6} - \tfrac{1}{8} + \tfrac{1}{10} - \tfrac{1}{12} + \ldots$, and this in turn is simply

$$\tfrac{1}{2}\left(1 - \tfrac{1}{2} + \tfrac{1}{3} - \tfrac{1}{4} + \tfrac{1}{5} - \tfrac{1}{6} + \ldots\right).$$

So, by taking the terms in a different order, we appear to have *halved the sum of the series*!

"HERE'S WHERE YOU
MADE YOUR MISTAKE."

And this is, indeed what has happened. If you have a finite collection of terms, then the order in which you add them doesn't matter, but we made the mistake of assuming the same was true for an infinite series.

We get some insight into the source of the difficulty if we take things to extremes. Suppose we decide to take *all* the positive terms first:

$$1 + \tfrac{1}{3} + \tfrac{1}{5} + \tfrac{1}{7} + \tfrac{1}{9} + \dots$$

the idea being to deal with the negative ones later. The trouble is, this series of purely positive terms doesn't, in fact, converge; its 'sum' is infinite, and a similar difficulty arises with the series of purely negative terms.

Even these insights, however, scarcely prepare us for an astonishing result obtained by Riemann in 1854: the sum of the terms $1, -\frac{1}{2}, +\frac{1}{3}, -\frac{1}{4}, +\frac{1}{5}\ldots$ can be made to converge to *any number we like* by adding the terms in a sufficiently cunning order.

---

So, when trying to solve a problem in mathematics we have to watch out for subtle mistakes, otherwise, we can easily get the wrong solution.

And there is something still more treacherous that occasionally lies in wait for the working mathematician. It may sometimes happen that, contrary to all expectation, the problem in question has *no solution at all*.

One interesting example of this is known as *Kakeya's problem*, after the Japanese mathematician who first posed it, in 1917. The problem is: find the smallest region in which a needle of unit length can be reversed, i.e. manoeuvred so that it rotates completely through 180°.

This seems an innocent enough problem, and the most obvious possibility that leaps to mind, I suppose, is a circle of radius $\frac{1}{2}$, because we can then just turn the needle around without any fuss at all. This circle has area $\frac{\pi}{4} \approx 0.78$.

Circle

Triangle

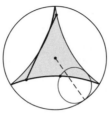

Hypocycloid

But a bit of thought shows that we can do better with an equilateral triangle of height 1. A little cunning is now needed: the 'trick' is to jam the needle into one corner, rotate it by 60° about that corner, then slide it into the next one, and so on. And the area now needed is only $\frac{1}{\sqrt{3}} \approx 0.58$.

A better choice still, however, is the so-called *hypocycloid* curve traced out by a point on the rim of a wheel of radius $\frac{1}{4}$ as it rolls inside a circle of radius $\frac{3}{4}$. It turns out that the area is then only $\frac{\pi}{8} \approx 0.39$, and the needle can still be rotated fully through 180° by a sort of 'three-point turn'.

For some years, in fact, it was believed that this was *the* solution to Kakeya's problem, and that no region of smaller area could be found.

Then, in 1927, A.S. Besicovitch dropped a bombshell by showing that there is in fact no solution at all, because we can make the area in which the needle turns *as small as we like*, provided we construct the region with sufficient cunning. The smaller the total area, the more the region in question takes on a very 'wispy' appearance, with many thin tendrils extending from its central region.

Some idea of the way in which this comes about can be obtained by imagining an equilateral triangle divided up into several pieces which are then slid together so that they overlap a little. It turns out that by dividing the

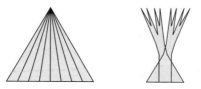

triangle into sufficiently many thin strips we can make the area of the resulting figure, called a Perron tree, as small as we like, and that by fitting several of these trees together we can arrange for the needle to be rotated completely, through 180°.

# CHAPTER TWELVE

# *What is the Secret of All Life?*

When I started school in the 1950s we had one elderly
teacher – I will call her Miss H – who had a real air of
mystery about her. She spoke in a thick foreign accent,
and gave her lessons from a darkened corner of the
classroom, where she sat huddled in a shawl.

Miss H taught us something which would now be
called biology, I suppose, and the way she did this was to
set exactly the same test, week after week, so that even
today I can still remember some of it.

The first question was: 'How many legs has an insect?', and the second was 'How many legs has a spider?'

Easy enough, you might say, and we soon got the hang of it.

But the very last question, number 23, was of a quite different character, and hardly suitable, really, for a group of very young children.

Even at the time, in fact, we had some faint notion of how deep this question was, and it seemed to have almost sinister overtones as it emerged from the darkness.

The question was:

*'What is the secret of all life?'*

According to Miss H, the answer was *chlorophyll*, though I don't think any of us really believed it, even then.

**chlorophyll** *or U.S.* **chlorophyl** ('klɔːrəfɪl) *n.* the green pigment of plants, occurring in chloroplasts, that traps the energy of sunlight for photosynthesis and exists in several forms, the most abundant being **chlorophyll a** ($C_{55}H_{72}O_5N_4Mg$): used as a colouring agent in medicines or food (**E140**). —'chloro₁phylloid *adj.* —₁chloro'phyllous *adj.*

Fifty years later, I don't feel any closer to knowing the real secret of all life; in fact, to be quite honest, I'm not entirely certain what Question 23 means any more. But if there really is a *scientific* 'secret' out there, I tend to assume that it lies somewhere in the most fundamental physical theories – such as quantum mechanics – that underpin not only biology but everything else as well.

And underpinning those physical theories, in turn, is mathematics, with one particular branch of the subject taking centre stage.

That branch is: *differential equations.*

---

Differential equations are statements, essentially, of how a system will change after a short period of time. They tell us, in other words, how *different* a system will be a short time later.

As an example, I should like to bring back the toy spider which we met on p. 98. For reasons I have never

understood, this particular spider has six legs, though I don't intend to pay them much attention in writing down the equations of motion.

And after we apply the basic principles of physics, these equations turn out to be *differential* equations.

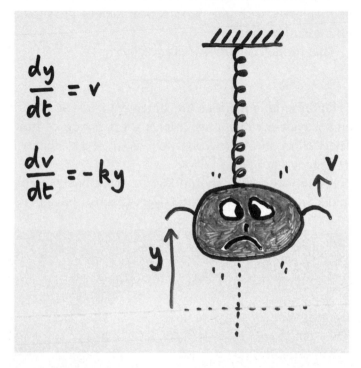

$$\frac{dy}{dt} = v$$

$$\frac{dv}{dt} = -ky$$

The two 'unknowns' here are $y$, the height of the spider above its equilibrium position and $v$, the spider's upward velocity. The constant number $k$, on the other hand, is to be regarded as 'given', and is essentially the strength of the spring divided by the mass of the spider.

To see what the nature of the problem is, compare all this with the elementary calculus of Chapter 6. There, we were given how some quantity $y$ depended on time $t$, and we saw how to deduce the *rate* at which $y$ changed with $t$, which we denoted by $\frac{dy}{dt}$.

But here, as with other differential equations, it is all the other way round. In the equations of motion for the spider we have some rather obscure information about the rates of change $\frac{dy}{dt}$ and $\frac{dv}{dt}$, and our task is to deduce how $y$ and $v$ depend on time $t$.

───────────

One way of attacking the problem is to use a computer.

To see the general idea, note first that at time $t$ the spider is at height $y$ and has upward velocity $v$. And a little later, at time $t + \delta t$, its height and upward velocity will have changed slightly, to $y + \delta y$ and $v + \delta v$.

Suppose, then, that we approximate $\frac{dy}{dt}$ by $\frac{\delta y}{\delta t}$ (see p. 55), and treat $\frac{dv}{dt}$ in a similar way. We can then convert the differential equations into

$$\delta y = v \times \delta t,$$

$$\delta v = -ky \times \delta t.$$

These are formulae for the small changes in $y$ and $v$ that occur after a small increase in time of amount $\delta t$.

And, most importantly, they allow us to write the 'new' values of $y$ and $v$ (i.e. $y + \delta y$ and $v + \delta v$) in terms of the 'old' ones:

$$\text{'new'} \, y = y + v \times \delta t,$$

$$\text{'new'} \, v = v - ky \times \delta t.$$

To put these rules into action on a computer we must first choose a value for $k$ (we will take $k = 1$ for simplicity) and we then need to choose a *small* value for the time step $\delta t$. We must also specify initial values for $y$ and $v$. The computer then uses the rules to calculate new values of $y$ and $v$, i.e. the values after one time step. It then takes *those* values of $y$ and $v$ and *repeats* the process to obtain values for $y$ and $v$ after two time steps, and so on.

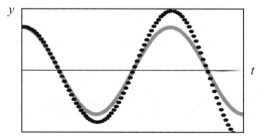

A typical outcome, with $\delta t = 0.1$, is shown above, and compared with how the spider actually moves up and down. The computer 'solution' to the problem clearly captures the oscillations quite well, but also displays a gradual – and incorrect – growth in the oscillation with time. This happens because the time step $\delta t = 0.1$ is not all that small, and the accuracy of the computer solution increases substantially if we use, say, $\delta t = 0.01$ instead.

In other words, lots of very small time steps provide the key to success when 'solving' differential equations on a computer.

And the general approach we have just described has wide applications, the daily weather forecast being just one example. TV bulletins are often sprinkled with references to 'the computer', and this typically means exactly the procedure above: use the laws of physics to write down the differential equations governing the motion of the atmosphere, convert them into 'updating' formulae to be applied after each small time step, and then get the computer to do these updatings, over and over again, until all the little time steps add up to the desired forecast time.

———————

Differential equations provide, then, some of the deepest links between mathematics and the physical world.

This first became apparent in the early eighteenth century, largely at the hands of Euler and his contemporaries. They used differential equations to greatly extend Newton's work on planetary motion, but found that these ideas opened up completely new areas as well, such as the dynamics of fluid motion.

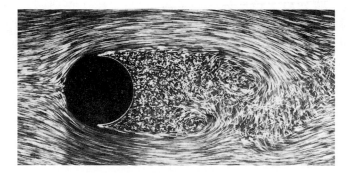

Later, in the nineteenth century, the whole subject of electromagnetism was transformed as soon as the differential equations were discovered, and the same is true of some of the greatest discoveries of the twentieth century, such as quantum mechanics.

And if – as I have heard some people claim – the great advances of the twenty-first century are to be in biology, then it seems to me likely that the associated mathematics will again involve differential equations in a major way.

There are, in fact, some signs that this is happening already. One particularly interesting example is a theory by J.D. Murray for the origin of animal coat markings. The basic question is, so to speak, how did the leopard *really* get its spots? And in the mathematical formulation of this theory, differential equations are, once again, at the heart of the whole matter.

I wonder what Miss H would have made of it all?

CHAPTER THIRTEEN

# e = *2.718*...

Imagine that you lend someone some money – say, one pound. (I hope this is not too implausible.)

Suppose, too, that by a combination of charm and guile you manage to persuade the borrower to agree an interest rate of 100% per annum. Then after one year you will collect $1 + 1 = 2$ pounds.

But if you are even more cunning (or degenerate) you might try to persuade the borrower to agree a compound interest rate of only 50% every six months. It sounds much the same – half the rate, payable twice as often – but it isn't. After six months you get $1 + \frac{1}{2}$ pounds, and after another six months you get $1 + \frac{1}{2}$ times *that*, i.e. $(1 + \frac{1}{2})^2 = 2.25$ pounds.

In the same way, an interest rate of $33\frac{1}{3}\%$ payable three times a year gives a return of $(1 + \frac{1}{3})^3 = 2.37$, which is a little better still.

So is there, perhaps, a fortune to be made this way?

| $n$ | $(1 + \frac{1}{n})^n$ |
|---|---|
| 1 | 2 |
| 10 | 2.59374 |
| 100 | 2.70481 |
| 1000 | 2.71692 |
| 10,000 | 2.71815 |
| 100,000 | 2.71827 |
| 1,000,000 | 2.71828 |

Well, not exactly, no.

This is because as we keep increasing the value of $n$ indefinitely, the quantity $(1 + \frac{1}{n})^n$ in fact tends to a finite *limit*.

And that limit is

$$e = 2.71828\ 18284\ 59\ldots$$

This strange number e pops up – like $\pi$ – in all sorts of different places in mathematics. And it arises, in particular, in connection with a fundamental question involving the rate at which things change.

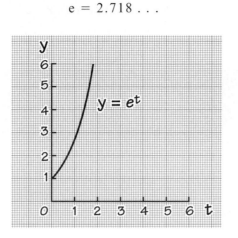

Now, if we have some quantity $y$ which depends on time $t$, its rate of change $\frac{dy}{dt}$ will usually be given by a different expression to that for $y$ itself. Thus if $y = t^2$ then $dy/dt = 2t$, if $y = \sin t$ then $dy/dt = \cos t$, and so on.

The question arises, then: is there any quantity $y$ *whose rate of change is always equal to $y$ itself*?

And the answer is yes, there is:

$$y = e^t.$$

This so called *exponential growth* is very rapid. Initially, at $t = 0$, the value of $y$ is 1, but by $t = 1$ this has multiplied by a factor $e = 2.718...$ And by $t = 2$ it has multiplied by a factor $e = 2.718...$ *again*, and so on. (When $t$ is a fraction, as opposed to a whole number, we may deduce the meaning of $e^t$ from the rule $e^a \times e^b = e^{a+b}$. Thus $e^{1/2}$ is the *square root* of e, for instance, because $e^{1/2} \times e^{1/2} = e^1 = e$.)

So, the particular quantity $e^t$ is such that

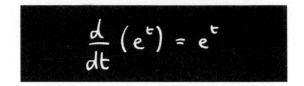

$$\frac{d}{dt}(e^t) = e^t$$

And it is, arguably, this property, above all else, that singles out e = 2.718 . . . as a very special number in mathematics.

---

One practical application of this result arises in a simple model for the spread of a disease.

Let $p$ be the fraction of the population which has the disease at time $t$, and suppose that the rate of infection is proportional to the percentage of the population which has the disease already. Then $dp/dt$ will be proportional to $p$, i.e.

$$\frac{dp}{dt} = cp,$$

where $c$ is a constant positive number.

Now, the solution to this differential equation turns out to be $p = p_0 e^{ct}$, where $p_0$ is the percentage of the population that has the disease initially, at $t = 0$.

"BUT THIS IS THE SIMPLIFIED VERSION
FOR THE GENERAL PUBLIC."

In this way, then, e enters the picture. And the picture
– according to this highly over-simplified model – is a
gloomy one, with more and more of the population
becoming infected at a faster and faster rate.

Another common way in which e arises in practice is through the whole idea of *instability*.

Imagine, for instance, a small droplet descending into a bowl of milk and causing a splash. Initially, the droplet is more or less spherical as it descends, and, given this initial symmetry, one might expect the milk surface to respond in a symmetric way.

And so it does – at first: the surface initially rises up from the point of impact in a smooth, circular ring with a thin wall which curves gently outwards.

Within a short time, however, and for no apparent reason, the top of this thin wall develops a wavy form, with 'crests' and 'troughs'. This waviness increases rapidly, and the crests, in particular, develop into highly elongated spikes, each of which eventually ejects a tiny droplet of milk.

Where, then, does this spiked pattern come from? Why doesn't the wall of milk remain symmetrical as it rises, with a perfectly circular form?

The answer is that the symmetrical motion is indeed possible in principle, but we never actually see it in practice, because tiny, unwanted disturbances – inevitable in any real experiment – grow rapidly with time and lead to a quite different outcome.

In short, the original, symmetrical motion is *unstable*; it is rather like trying to balance a pencil on its point. And, most significantly for present purposes, in the very early stages of the instability the small disturbances grow at a rate proportional to the current size of the disturbances themselves, and in this way the number

$$e = 2.718 \ldots$$

enters the picture.

Another remarkable example of this general kind can be seen in the photographs below. This is instability in a chemical context – the so-called Belousov–Zhabotinskii reaction, discovered in 1951. In this case, small spatial non-uniformities in the concentrations of the various chemicals grow with time, and over the course of a few minutes these organize themselves into a striking and regular pattern, this time in the form of spiral waves.

But you don't need bowls of milk or lots of exotic chemicals to see the number e = 2.718 . . . at work.

Take two packs of ordinary playing cards, shuffle them, and place them side by side, face down. Now turn over one card from each pack. The chances that the two cards will be the same are, of course, very slim.

But suppose now that we keep on doing this, turning cards, over, in pairs, looking for a 'match'. Are we likely to get through *all 52 pairs* of cards without encountering a matching pair?

The answer, in fact, is . . . . no, we're not. The reason for this is that

$$\textit{probability of no matching pair} \approx \frac{1}{e}$$

and this probability is less than $\frac{1}{2}$, i.e. less than 50%.

---

We end this chapter, however, with a remarkable representation of the quantity $e^t$ as an infinite series:

$$e^t = 1 + t + \frac{t^2}{2} + \frac{t^3}{2 \times 3} + \frac{t^4}{2 \times 3 \times 4} + \cdots$$

And we don't have to take this strange result on trust either; it is relatively easy to check that it is consistent with the key result noted earlier, namely

$$\frac{d}{dt}(e^t) = e^t.$$

To do this, we simply use the table on p. 58 to calculate the rate of change of each individual term of the infinite series.

In this way, then,

$$\frac{d}{dt}(e^t) = 0 + 1 + \frac{2t}{2} + \frac{3t^2}{2 \times 3} + \frac{4t^3}{2 \times 3 \times 4} + \ldots$$

and after some cancellation the right-hand side becomes $1 + t + \frac{t^2}{2} + \frac{t^3}{2 \times 3} + \ldots$, which is just the original series representation for $e^t$.

Another simple thing we can do is substitute the particular value $t = 1$ in the original series, for this gives an elegant representation for the number e itself:

$$e = 1 + 1 + \frac{1}{2} + \frac{1}{2 \times 3} + \frac{1}{2 \times 3 \times 4} + \ldots$$

Yet even this is nothing compared with the fate that we have in mind for e at the very end of the book, where it will take its place in what many regard as the most wonderful mathematical result of all time.

CHAPTER FOURTEEN

# *Chaos and Catastrophe*

A few years ago, the whole idea of *chaos* hit the headlines in a big way. It was going to be the answer to everything, and the subject became so fashionable that it even made a fleeting appearance in one of Steven Spielberg's block-buster films.

While it all got a bit out of hand, the key ideas are important, and, oddly enough, there was some indication of them, in the nineteenth century, in connection with what is now called the gravitational *three-body problem*.

Today, we can study the problem relatively easily using a computer. In the example below, for instance, three equal point masses attract one another according to the law of gravitation, starting at the positions marked by the numbers. So, masses 2 and 3 move towards one another, and then whirl each other around before separating again. Mass 2 then goes on to have a similar 'close encounter' with mass 1 instead.

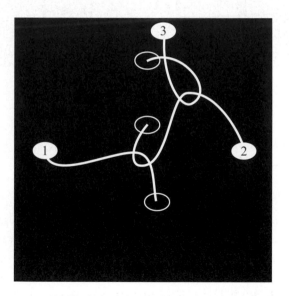

The next event in the sequence is that 2 and 3 whirl each other around, in a gentle way, as a result of which 3 is sent on to a closer encounter with 1. And this is followed, in the final frame below, by two *very* close encounters, first between 1 and 2 and then between 2 and 3.

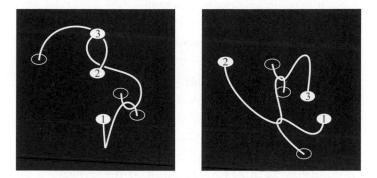

And because the outcome of each close encounter depends very acutely on *how* close the encounter is, the motion of the system as a whole depends strongly on the exact starting positions and starting velocities of the various masses.

This, then, is the essence of chaos: irregular, erratic motion which is extremely sensitive to the initial conditions. And despite this early work on the topic it was not until the 1960s – or even later – that mathematicians began to realize that this kind of behaviour could occur quite naturally in a large class of other systems.

And these systems weren't necessarily mechanical, either; some of the most interesting examples of chaos came from problems involving electronic circuits, while others came from applications in chemistry or even biology.

Suddenly, chaos was everywhere.

———————

All this – and particularly the sensitivity to initial conditions – came as very bad news to people engaged in long-term prediction. Weather forecasting, for instance, is notoriously difficult, and one reason for this may be that the atmosphere goes through chaotic phases, where tiny uncertainties about its initial state translate into enormous uncertainties about how it will be in the future.

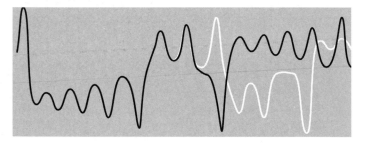

A hallmark of chaos: two almost imperceptibly different starting conditions lead to two completely different outcomes, within a relatively short space of time.

One of the early pioneers of chaotic dynamics, Ed Lorenz, made the point only too well in a landmark paper published in the *Journal of the Atmospheric Sciences* in 1961:

> . . . When our results . . . are applied to the atmosphere, . . . they indicate that prediction of the sufficiently distant future is impossible by any method, unless the present conditions are known exactly. In view of the inevitable inaccuracy and incompleteness of weather observations, precise very-long-range forecasting would seem to be non-existent.

But what really surprised many mathematicians in the 1970s was the way in which chaos could often be found in much 'simpler' systems.

A good example first arose in 1976 in connection with population dynamics; it is the rule

$$x_{n+1} = ax_n(1 - x_n)$$

for calculating each new number in the sequence $x_1$, $x_2$, $x_3$, . . . from the one before. Here $a$ is a fixed number, chosen in advance, between 0 and 4.

In other words, we choose a starting value $x_1$ (between 0 and 1), and then we use the above rule to calculate $x_2 = ax_1(1 - x_1)$. Knowing $x_2$, we then use the rule again to calculate $x_3 = ax_2(1 - x_2)$, and so on. It all seems 'simple' enough, and programming a computer to do the calculations is child's play.

Suppose, then, that we choose a value for $a$, choose a starting value $x_1$, and run the program. How does the number sequence $x_1, x_2, x_3, . . .$ develop? Well, if $a$ is less than 1, then $x_n$ just gets smaller and smaller as $n$ increases. If $a$ lies between 1 and 3, then, as $n$ increases, $x_n$ gradually converges to the constant value $1 - 1/a$ instead.

If $a$ is between 3 and 3.449, on the other hand, something more interesting happens: the system settles into a regular oscillation, with $x_n$ flipping to and fro between two different values as $n$ increases. And if $a$ is slightly larger still, more complicated regular oscillations occur.

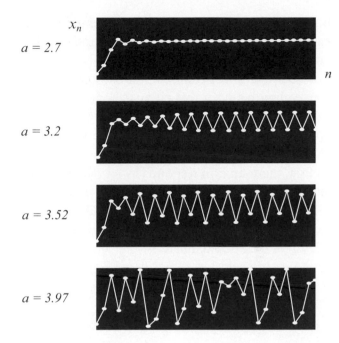

Most interesting of all, however, is the behaviour when *a* is greater than 3.570. Typically, the system doesn't then settle down to a steady value or any kind of regular oscillation. Instead, successive values of $x_n$ fluctuate in an apparently haphazard way, so that the oscillations never quite fall into a regular, repeating pattern.

Here, once again, is chaos, and in one of the 'simplest' mathematical systems that one could imagine.

But chaos isn't the only recurring theme in modern research in dynamics. There is another, quite different one as well.

Consider again, for example, the idea of a thin *soap film* extending between two circular hoops (chapter 7). Consider, in particular, what happens if we start with the hoops close together and then gradually pull them further apart.

At first, the soap film responds by *gradually* changing its shape; it becomes more curved, with a greater degree of 'pinching' in the middle, as one would expect. And this continues until the distance between the hoops is 0.6627 times the hoop diameter.

Yet as soon as the distance apart is increased beyond this critical value, the whole film suddenly collapses, leaving instead two separate, flat, films, one on each hoop. And this collapse happens quite without warning, so to speak, long before the central parts of the film show any signs of touching.

This general kind of behaviour, where a gradual change in some parameter can lead to a sudden and unexpected large change in the system as a whole, is known as a *catastrophe*.

And it's not at all difficult to find physical systems which can undergo catastrophic changes *and* chaotic motion. Indeed, the simplest one I know is called a *pendulum*.

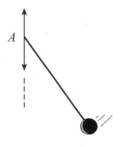

Left to its own devices, of course, a pendulum rod will just swing to and fro at a certain frequency until the oscillations are damped by friction. But I propose to liven things up by vibrating the pivot up and down, albeit in an entirely regular manner. And the most effective way of doing this – so it turns out – is to vibrate the pivot up and down at twice the pendulum's natural swinging frequency.

As usual, we begin by writing down the appropriate differential equations, which determine how the system changes from one moment to another. We then use a computer to 'solve' the equations, and some of the most interesting results emerge if we allow ourselves to gradually change the magnitude $A$ of the pivot motion while the pendulum is swinging.

When $A$ is very small, the pendulum always ends up hanging downward, in the way one might expect. But if we gradually increase $A$ beyond a certain critical value, the pivot vibrations are large enough to make this downward-hanging state unstable, and the pendulum settles instead into a symmetric swinging oscillation (a).

Then, if we increase $A$ further this oscillation itself eventually becomes unstable, and an *un*symmetric oscillation takes its place, with the pendulum swinging higher on one side of the vertical than the other (b).

At even higher values of $A$, the pendulum moves chaotically, with a mixture of irregular swinging oscillations and intermittent, half-hearted 'whirling' motions about the pivot (c).

Yet, as we gradually increase $A$ still further, something remarkable happens: the chaotic motion suddenly collapses completely, and the pendulum switches instead to a vigorous but completely regular whirling motion about the pivot (d).

This, then, is a catastrophe, and like other catastrophes it cannot immediately be 'undone'. For if we straight

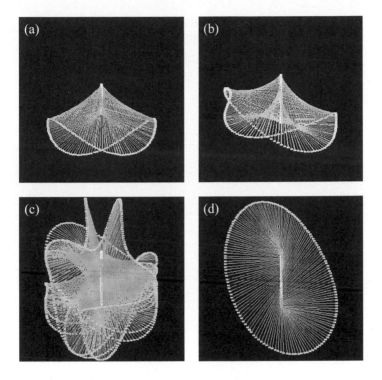

away start *decreasing A* again, this regular whirling motion does not instantly revert to the chaotic motion that preceded it. Instead, the whirling motion persists, though with diminishing vigour as *A* is reduced.

It is only when *A* decreases below some substantially smaller critical value that the regular whirling motion suddenly collapses, in another catastrophe, and the system jumps back to a regular swinging oscillation.

Not surprisingly, all this exotic behaviour is best seen, really, using computer animations rather than still pictures. Readers who would like to do this can find the animations on the *1089 and All That* website (see p. 172).

CHAPTER FIFTEEN

# *Not Quite the Indian Rope Trick*

Mathematics pops up in the most peculiar places. For instance, you wouldn't think it had much to do with the *Indian Rope Trick*, would you?

This is, perhaps, the most famous trick in the whole history of magic: a length of rope is thrown up into the air, and instead of falling back down again it stays up there, defying gravity. A small child then climbs up the rope, and disappears at the top.

Accounts of this date back to the fourteenth century, though the trick itself has been most elusive, at least in its proper form, meaning in the open air and in broad daylight.

Yet there have been some remarkable attempts quite recently. One of these happened in February 1999, at a conference on magic in Cochin, southern India. In the middle of one of the main streets, the magician, Professor Padmarajan, played on a snake charmer's pipe as a length of rope emerged from a large basket and rose stiffly into the air, to a height of about 15 feet. A five-year old child then climbed to the top.

By all accounts, it was very impressive.

But what on earth has it got to do with mathematics?

And why did a well-known British newspaper then phone *me*, in Oxford, to ask if I knew how the trick was done?

———————

To answer this, we need to turn back, believe it or not, to 1738, when the Swiss mathematician Daniel Bernoulli published a novel paper on pendulum motion.

Now, when a single pendulum swings to and fro by a small amount, it performs a certain number of oscillations in one unit of time, and this is called the natural frequency of the pendulum.

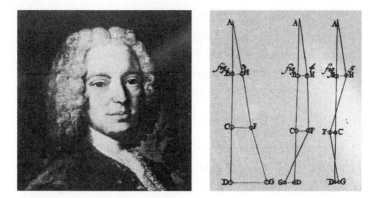

The 3 modes of oscillation of a triple pendulum, from
Daniel Bernoulli's original paper of 1738.

But Bernoulli was interested in *multiple* pendulums,
i.e. $N$ pendulums suspended from one another, so as to
form an $N$-linked hanging chain. And he discovered that
this system can oscillate at any one of $N$ different natural
frequencies

$$f_1, \ldots, f_N,$$

where $f_1$ denotes the smallest and $f_N$ the largest. In the
lowest-frequency mode the pendulums swing to and fro
more or less together, much as if they formed one long,
single pendulum. In the highest-frequency mode, on
the other hand, adjacent pendulums swing in opposite
directions at any given moment.

That is how things were, then, in 1738. And in order to make some kind of connection with the Indian Rope Trick we now need to take Daniel Bernoulli's system of *N* linked pendulums and give it a strange new twist.

What we need to do, in fact, is turn it *upside-down*.

---

So it was that one rainy November afternoon in 1992 I found myself proving a strange new theorem.

And while I was quite used to getting surprising things out of differential equations, I have to say that this one had me shaking my head in disbelief. For, according to the theorem, it is possible to take *N* linked pendulums, turn then upside-down, so that they are all precariously balanced on top of one another, and then *stabilize them in this position by vibrating the pivot up and down*.

This is very different, incidentally, from balancing an upturned pole on the palm of your hand. There, you move your hand from side to side in response to how the pole happens to be falling over at the time. Here, on the other hand, the pivot vibrations are strictly up and down, and completely regular, so that they continue in a predetermined way, quite unchanged, as the upturned pendulums wobble about.

Now, the idea that a single upside-down pendulum can be stabilized in this way is quite well known, and dates from a remarkable paper in 1908 by Andrew Stephenson, an applied mathematician at Manchester

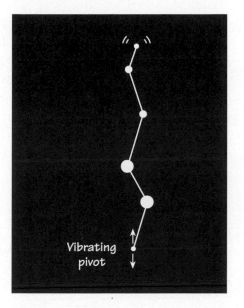

Vibrating
pivot

University. The new theorem is in the same spirit, then, but goes much further, and says that the same feat can be performed with *any finite number* of linked pendulums, no matter what their various shapes and sizes.

And one particularly pleasing aspect – in my view, at least – is that the theorem contains a direct link with Bernoulli's work of 1738. For instead of involving explicitly all sorts of messy and cumbersome details of the particular pendulum system in question, it involves just the two key numbers $f_1$ and $f_N$, which, as we have seen, relate to swinging oscillations of the system about the usual, downward-hanging state.

The upside-down pendulum theorem.

In practice, $f_N^2$ is usually much greater than $f_1^2$, and the theorem then states that the pivot vibrations will stabilize the upside-down state if two conditions are met, namely, the ones on the blackboard above.

There *a* denotes half the total distance through which the pivot moves up and down, $f_\text{p}$ denotes the frequency of the pivot vibration, and *g* is the acceleration (9.81 m sec$^{-2}$) of any object that falls freely under gravity.

The upshot of all this, then, is that the 'trick' can always be done, so long as the pivot is vibrated up and down by a *small enough amount* and at a *high enough frequency*.

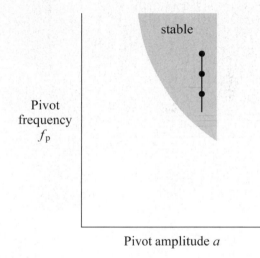

Pivot amplitude *a*

The mathematics seemed clear. I even constructed a computer simulation, based directly on the governing differential equations, and it seemed to fully support the theorem. But could it possibly work out in practice?

In this respect I was very lucky in having Tom Mullin nearby, in Oxford, at the Clarendon Laboratory.*

Tom is one of the world's leading experimentalists on chaos, and I had already seen him use multiple pendulums, with a vibrating pivot, to demonstrate chaotic motion. In order to test the predictions of my theorem I needed to persuade him to try some new experiments at a much higher vibration frequency.

In the event, he didn't take much persuading. It took him just three days to get the upside-down double pendulum working, and, though the triple pendulum proved rather more problematic, he eventually got that working too. The frequencies involved did then turn out to be quite high: with a 50 cm inverted triple pendulum, for instance, the pivot was vibrating up and down through 2 cm or so at a frequency of about 40 cycles per second.

Nonetheless, the 'trick' really did work, and it worked, in fact, far better than we could ever have imagined. We were quite taken aback by just *how* stable the inverted state could be, and provided the pendulums were kept roughly aligned with one another, we could push them over by as much as 40 degrees or so and they would still gradually wobble back to the upward vertical.

In November 1993 we published our results in *Nature*, and hoped, I think, that they would cause a fair amount

* Tom Mullin is now at the Department of Physics and Astronomy, University of Manchester.

of public interest. As *Nature* is published on Thursdays, and the science pages of the national press often run with a *Nature* story the following day, we waited expectantly till Friday morning. In the end, however, we were beaten by another article: 'Lifespan and testosterone', which was all about whether or not men live longer if they're castrated.*

In the months that followed, however, Tom and I gave occasional lectures on our work, and at that stage we both began to mention – however playfully – the loose analogy between our strange balancing act and the Indian Rope Trick. Largely as a result of this, I think, the BBC eventually got wind of our work, and in October 1995 we made a brief appearance, with the experiment, on the TV programme *Tomorrow's World*.

What was it that Andy Warhol once said? Something about everyone being famous for fifteen minutes? It was more like three or four minutes in our case, but it was still fun while it lasted.

———————

All that was some years ago, of course, and I'm now so used to the upside-down pendulum theorem that there are times when it seems almost 'normal'. Yet, deep

---

* In case any readers should be wondering, the answer, apparently, is no, they don't.

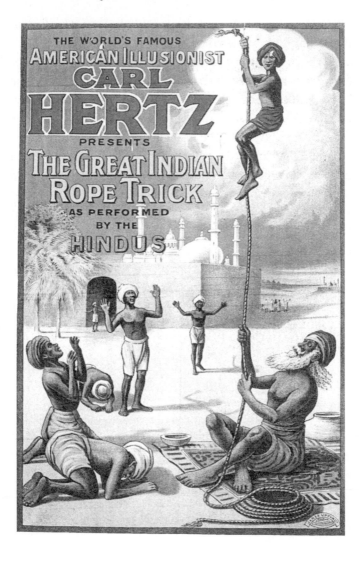

down, I know that it isn't. It may not be quite the Indian Rope Trick, but it's still pretty weird.

Perhaps the best proof of this came the day after the TV programme, when the BBC received a telephone call of the 'disgusted of Tunbridge Wells' variety. The caller claimed that our balancing act was obviously impossible, and contrary to the laws of physics, and he seemed genuinely upset with *Tomorrow's World* for 'lowering their usually high standards' and 'falling prey to two tricksters from Oxford'.

# CHAPTER SIXTEEN

# Real or Imaginary?

Once in a while, someone conducts a poll by asking a bunch of mathematicians what they reckon to be the ten most wonderful pieces of mathematics of all time. And while some of the answers depend a bit on when the poll is taken, and who exactly is asked, one particular result *always* wins.

The result in question involves the square root of $-1$, i.e.

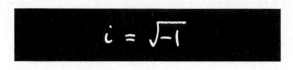

$$i = \sqrt{-1}$$

and the first thing we must do in this final chapter is get to grips with this mysterious quantity.

After all, what kind of number *is* it that, when multiplied by itself, gives $-1$? No positive number has this property, and no negative number, either; all such 'real' numbers, when squared, give a positive number.

For precisely this reason, we call $i = \sqrt{-1}$ an *imaginary* number, though this scarcely explains how it came to be taken at all seriously, let alone why it is now used daily, in the most down-to-earth of calculations, and almost without thinking, by engineers and scientists all over the world.

It is sometimes assumed, in fact, that imaginary numbers first took their place in mathematics through quadratic equations like $x^2 + 1 = 0$, which can be re-cast as $x^2 = -1$. But while mathematicians could have said that the solutions of this equation were $x = i$ or $x = -i$, on the whole, they didn't. It seemed more sensible – more honest, even – to just say that the equation had no solutions at all.

So how on earth did such a strange entity as $i = \sqrt{-1}$ come to be taken seriously?

Interestingly, it wasn't quadratic equations which propelled imaginary numbers into the limelight, but *cubic* equations like

$$x^3 = 15x + 4.$$

Now, the sixteenth-century Italian scholar Girolamo Cardano obtained a general formula for the solutions of cubic equations of this type, and in this particular case his formula gives

$$x = \sqrt[3]{2 + 11i} + \sqrt[3]{2 - 11i},$$

where $\sqrt[3]{\phantom{x}}$ denotes a cube root.

If we adopt the same hard-nosed attitude to imaginary numbers, and say that $i = \sqrt{-1}$ 'doesn't exist', we have to regard the above equation as nonsense, and we are driven to the conclusion that the cubic equation in question has no solutions – or, at the very least, no 'real' solutions.

But there is one big difficulty with this – there clearly *is* a real solution to the cubic equation $x^3 = 15x + 4$, namely

$$x = 4,$$

because this particular value of $x$ makes both sides of the equation equal to 64.

We are left with the problem, then, of how to extract the simple answer $x = 4$ from Cardano's supposedly general formula involving cube roots and imaginary numbers.

This problem was solved by Raffaele Bombelli in his influential treatise *L'Algebra*, published in 1572. Bombelli proceeded by treating the imaginary number $i = \sqrt{-1}$ seriously, and he manipulated it according to all the usual rules of algebra, just as if it were a 'real' number.

In particular, he noticed the following. If we take $2 + i$ and multiply it by itself we get $(2 + i)^2 = 4 + 4i + i^2$. And as $i^2 = -1$ this can be rewritten as $3 + 4i$. So, if we

Extract from Bombelli's book of 1572. He writes p for plus and m for minus, and – allowing for further differences in notation – we can pick out 11i in the form $\sqrt{0 - 121}$.

now multiply by 2 + i again, so as to form $(2 + i)^3$, we get $(3 + 4i)(2 + i) = 6 + 11i + 4i^2$, which can be rewritten as $2 + 11i$.

So

$$(2 + i)^3 = 2 + 11i,$$

and, in a similar way,

$$(2 - i)^3 = 2 - 11i.$$

And the significance of this in connection with the equation

$$x^3 = 15x + 4$$

is, of course, that it allowed Bombelli to interpret Cardano's solution

$$x = \sqrt[3]{2 + 11i} + \sqrt[3]{2 - 11i}$$

as

$$x = 2 + i + (2 - i)$$
$$= 4,$$

so producing the 'real' solution previously observed.

It was in this way, then, by resolving a serious paradox concerning *cubic* equations, that $i = \sqrt{-1}$ made its way into the mainstream of mathematics.

But it wasn't until much later that imaginary numbers really began to flourish, and no book was more influential in this respect than Euler's ground-breaking calculus text *Introductio in analysin infinitorum*, published in 1748.

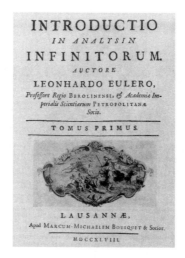

One astonishing result, in particular, put $i = \sqrt{-1}$ on the map, and in order to see this we need to recall from Chapter 10 the quantities $\sin \theta$ and $\cos \theta$, which are all about oscillations.

Now, it had been known since Newton's time that these quantities can be represented in the form of infinite series, one containing only odd powers of $\theta$ and one containing only even powers:

$$\sin \theta = \theta - \frac{\theta^3}{2 \times 3} + \frac{\theta^5}{2 \times 3 \times 4 \times 5} - \ldots$$

$$\cos \theta = 1 - \frac{\theta^2}{2} + \frac{\theta^4}{2 \times 3 \times 4} - \ldots$$

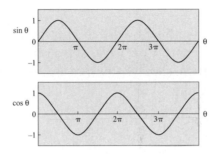

(It is quite easy to check, for instance, that these series representations are consistent with the two key results from p. 97.)

Next, recall from Chapter 13 the number e = 2.718..., which is all about compound interest payments, or playing cards, or instabilities of milk-drop splashing. Recall, in particular, from p. 132, this infinite series:

$$e^t = 1 + t + \frac{t^2}{2} + \frac{t^3}{2 \times 3} + \frac{t^4}{2 \times 3 \times 4} + \ldots$$

which turns out to be valid for all real numbers $t$.

At this stage, then, we have some relatively simple, attractive infinite series representations for sin θ, cos θ, and $e^t$. And now we are ready to take a bold step.

Some might even call it reckless.

Some, in fact, would call it absolutely outrageous.

---

Before we take this 'bold step', however, let us look back for a moment. For, while it is always possible to go further, this particular journey into mathematics is coming to an end.

Back at the beginning I was brave enough (some might say foolish enough) to sum up mathematics in just six words:

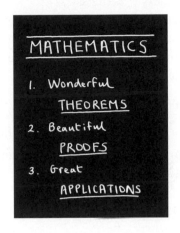

**MATHEMATICS**

1. Wonderful <u>THEOREMS</u>
2. Beautiful <u>PROOFS</u>
3. Great <u>APPLICATIONS</u>

I hope you feel that this strikes the right note. I hope, even, that you may have your own favourite theorems, methods of proof, and so on. Some of the strange results concerning $\pi$ in Chapter 9, for instance? Or the whole idea of proof by contradiction? Some of the applications, perhaps, like planetary motion and chaos? Or maybe the sheer 'human element' of Chapter 11, with mathematicians actually getting things badly *wrong*?

Or maybe even the '1089' trick, back on page 1? I wouldn't want to dismiss even that; it was *my* favourite too, many years ago.

But it is now time to end.

So let me thank you for making this journey with me, and turn attention to our one piece of unfinished – and thoroughly reckless – business.

Because what we now do is take the series representation of $e^t$, and substitute in – absolutely shamelessly – the *imaginary* quantity $t = i\theta$, where $\theta$ is a real number and $i = \sqrt{-1}$. This leads at once to

$$e^{i\theta} = 1 + i\theta - \frac{\theta^2}{2} - \frac{i\theta^3}{2 \times 3} + \frac{\theta^4}{2 \times 3 \times 4} + \frac{i\theta^5}{2 \times 3 \times 4 \times 5} \cdots$$

and if we separate out the real and imaginary terms on the right-hand side we obtain

$$e^{i\theta} = \left(1 - \frac{\theta^2}{2} + \frac{\theta^4}{2 \times 3 \times 4} - \cdots\right)$$
$$+ i\left(\theta - \frac{\theta^3}{2 \times 3} + \frac{\theta^5}{2 \times 3 \times 4 \times 5} - \cdots\right)$$

But the two infinite series in brackets here are *precisely the ones for* $\cos\theta$ *and* $\sin\theta$ on p. 165! And so we arrive at

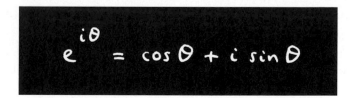

This extraordinary formula, due to Euler in 1748, provides a suitably high point on which to end the book.

Firstly, we have obtained it by putting together a whole variety of relatively sophisticated mathematical ideas, including calculus, infinite series and imaginary numbers.

Secondly, the formula is of great practical value; it is the sole reason, really, why virtually any engineering or physics book on oscillations has both e and $i = \sqrt{-1}$ all over the place, greatly simplifying many of the calculations.

And finally, by substituting in the special value $\theta = \pi$, and noting from p. 165 that $\sin \pi = 0$ and $\cos \pi = -1$, we obtain at last

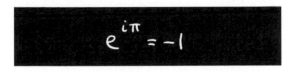

$$e^{i\pi} = -1$$

While any one of us is fully entitled – of course – to a quite different opinion, this amazing connection between e, i and $\pi$ is viewed by many mathematicians as, quite simply, the most stunning result in the whole subject ... ... *so far*.

# Suggestions for further reading

**From Calculus to Chaos**, by David Acheson, 1997, Oxford University Press. *Informal preview of applied mathematics at university level.*

**Numbers and Proofs**, by R. B. J. T. Allenby, 1997, Arnold. *Introduction to pure mathematics at university level.*

**What is Mathematics?** By Richard Courant and Herbert Robbins, 1941, Oxford University Press. *Classic introduction to the fundamentals of the subject, from an advanced viewpoint.*

**Mathematics: The New Golden Age**, by Keith Devlin, 1998, Penguin. *Lively account of some modern developments.*

**Journey through Genius: The Great Theorems of Mathematics**, by William Dunham, 1990, Wiley. *A handful of great theorems, with plenty of history.*

**Chaos**, by James Gleick, 1988, Heinemann. *Popular bestseller.*

**A Mathematician's Apology**, by G. H. Hardy, 1940, Cambridge University Press. *Classic, highly personal view of life as a pure mathematician.*

**The Parsimonious Universe**, by Stefan Hildebrandt and Anthony Tromba, 1996, Copernicus. *Beautifully-illustrated book on minimization problems.*

**Makers of Mathematics**, by Stuart Hollingdale, 1989, Penguin. *Concise and balanced history of the subject.*

**Fermat's Last Theorem**, by Simon Singh, 1997, Fourth Estate. *Best-selling account of how the theorem came to be proved at last.*

**From Here to Infinity**, by Ian Stewart, 1996, Oxford University Press. *Sweeping survey of modern developments.*

**The Penguin Dictionary of Curious and Interesting Geometry/Numbers**, by David Wells, 1991/3, Penguin. *Unbeatable for dipping-into on a dark, rainy afternoon.*

# The *1089 and All That* Website

This website, at **www.jesus.ox.ac.uk/~dacheson** contains various items relating to the book, including computer simulations of planetary motion (p. 45), properties of $\pi$ (Chapter 9), convergence of infinite series (p. 108), computer solution of differential equations (p. 118), the gravitational 3-body problem (p. 136), 'elementary chaos' (p. 141), chaotic pendulums (p. 145), and the upside-down pendulum theorem (p. 152).

# Acknowledgments

This book has been one of the most ambitious and imaginative projects I've ever undertaken. It bears little resemblance to the book I innocently started writing five years ago, and many people have helped me along the way.

I should like to single out for special mention my father, John Acheson, and my good friend and fellow guitarist Don Fowler. It is a great sadness to me that neither lived to see the book in its final form.

Special thanks are also due to Robert Acheson, Sophia Fowler and Janet Mills.

But I had encouragement and advice from many other people too, including Beth Ashfield, Arthur Barnes, Joyce Batty, Bertie Bellis, Viv Bowyer, Fyfe Bygrave, Peter Clifford, Maggie Couldwell, David Crawford, Andrew Dancer, John Gittins, Raymond Hide, Tony Hubbard, David Hughes, Andy Hunt, John Ireland, Michael Mesterton-Gibbons, Tom Mullin, Anthony Pilkington, John Roe, Chris Simmonds, Viktor Thaller, Sandra Tinson and John Wilson.

Lastly, I wish to thank Richard Lawrence, Alison Jones, and other members of Oxford University Press, for all their support, patience and good humour.

## Picture credits

1,3: From the *I-SPY Annual* for 1956, News Chronicle. (By permission of Michelin, ref B/1101.)

2, 159 (*adapted*): The Robert Opie Collection.

4: By kind permission of Glen Baxter (From *Glen Baxter; His Life*, Fontana-Collins 1983.)

24, 59, 109, 127: © Sidney Harris 2002.

27: From *The Guardian* for 24 June 1993. © *The Guardian*.

29, 113, 139, 166: From Giles cartoons which first appeared in the *Daily Express* on, respectively, 14 March 1963, 14 January 1954, 28 February 1946, 16 September 1976. © Express Newspapers.

30: Copyright © Ronald Searle 1953, by kind permission of Ronald Searle and the Sayle Agency.

32: From the *Gamages* Magic Catalogue, London ~ 1950.

36: Popperfoto.

41: From *The Modern World Book of Hobbies*, Sampson Low, Marston & Co., London ~ 1950.

53, 128: Harold & Esther Edgerton Foundation, 2002, courtesy of Palm Press, Inc.

61: © Mirror Syndication International.

71: By kind permission of C. Isenberg.

73: *On Early Shift* (Greenwood Signal Box, New Barnet) by Terence Cuneo, National Railway Museum/Science and Society Picture Library.

93: Django Reinhardt in Manchester, England, 1938. © Duncan Schiedt Collection.

# *Index*

algebra 29
  complaints about 29, 31
  and geometry 37
  made 'interesting' 4
  use of 32, 33, 35, 57, 78, 80,
    85, 162
angle 12, 63, 72, 95, 111
area
  of circle 13, 84
  of general region 76
  of triangle 11, 85
  of rectangle 11
axes, coordinate 37

bead, sliding 67
Bernoulli, D. 148
Bernoulli, J. 67
biology, mathematical 121
Bombelli, R. 162

calculus 53
  in action 56, 64, 97, 116, 126,
    132
  changes, small 54
  $dy/dt$ 54
  idea 53, 60
  and pi 88
  of variations 67

Cardano, G. 161
cards, playing 131
catastrophe 142, 145
change, rate of, *see* rate of
  change
change, small 54
chaos 135
  pendulum 143
  in 'simple' system 140
  three-body problem 136
chemical oscillations 130
chlorophyll 114
circle
  area of 13, 85
  circumference of 13, 83, 92
  divisions of 17, 85
  and Kakeya's problem 111
  and the odd numbers 14, 89
  packing problem 104
cisterns, filling 5
coins, tossing 92
compound interest 123
computer solutions 117
cone 44
conjecture, wrong 107
conjuring, mathematical 1, 32
contradiction, proof by, *see*
  proof

convergence 74
coordinates 37
cos θ 96, 165, 168
counter-example 107
cowboy problem 62
cubic equation 161
cycloid 68
cylinder 86

Descartes, R. 37, 39
diameter 84
differential equations 116, 126, 144, 150, 153
    computer solutions 117
differentiation 58
disease, spreading 126
d$y$/d$t$
    meaning 54, 60
    examples 57, 58, 97
    *see also* rate of change

e 123, 133, 169
e$^t$ 126, 132
electromagnetism 121
ellipse 43
equations
    cubic 161
    and curves 38
    differential 116
    quadratic 35
Euclid 23
Euler, L. 20, 91, 107, 120, 164
equals sign 33
exponential growth 125

Fermat's last theorem 26
fluid dynamics 120
fractions 77

geometry 9
    link to algebra 37
    of circles 84
    ellipse 43
    Kakeya's problem 110
    packing problem 104
    Pythagoras's theorem 10
    shortest path problems 62, 70
    topology 15
gravitational force 47

Halley's comet 42
harmonics 100
Hooke, R. 47
hypotenuse 95

i 160
    connection with e and π 159, 169
    meaning 160
    origins 161
imaginary numbers 160
Indian rope trick 147
induction, proof by, *see* proof
infinite series
    convergence 74
    divergence 75
    and e 132
    and pi 14, 89, 90
    rearranged 108

instability  128
integer  77

Kakeya's problem  110
Kepler, J.  44
Königsberg bridge problem  20

Leibniz, G.  58
leopard spots  121
Lorenz, E.  139

Malfatti's problem  104
maximum  65
minimum problems  61, 66, 67,
    71
modes of oscillation  100
Molesworth, N.  30
Moon  36
Mullin, T.  155

network  70
Newton, Sir Isaac  49
nodes  101
number line  77
numbers
  imaginary  160
  irrational  77
  odd  14
  prime  22
  rational  77
  real  77, 161

oscillations
  frequency  148
  pendulum  148

sine curve  94, 99
spider on spring  98

packing  104
parabola  38
pendulums
  chaotic  143
  multiple  149
  oscillating  148
  upside-down theorem  150
pi ($\pi$)
  and circles  13, 83
  and infinite series  14, 90
  and probability  92
  Viete formula  88
  Wallis product  88
planetary motion  45
polygon  84
population models  140
prime numbers  22
*Principia*  51
probability  92, 132
proof
  importance of  17
  by contradiction  19, 78
  by induction  79
  of Fermat's Last Theorem
    27
  of Pythagoras's theorem  11
Pythagoras's theorem  10

quadratic equations  35

rate of change  54, 60
  of cos θ  97

rate of change (*contd.*)
  in differential equations  116
  of $e^t$  126, 133
  of $\sin \theta$  97
  of $t^2$  57
  of $t^n$  58
  zero at a maximum  65
rectangles  76
*reductio ad absurdum*, *see* proof
    by contradiction
Reinhardt, D.  93

series, infinite, *see* infinite
    series
Sherlock Holmes  19
shortest path  70
shortest time  67
sine curve  94, 100
$\sin \theta$  96, 165, 168
slope of curve  65
soap film  61, 66, 71, 142
speed  56

spider on spring  98, 115
splash, milk drop  128
spring, oscillating  98, 116
Stephenson, A.  150
square root
  of two  12, 70, 78, 88, 152
  of three  12, 72, 105, 111

three-body problem  136
topology  15
triangle
  area of  11, 85
  equilateral  105, 111
  right-angled  12, 95
trigonometry  95

upside-down pendulum
    theorem  152

Wallis's product  88
weather forecasting  119, 138
Wiles, A.  27